About Island Press

Island Press, a nonprofit organization, publishes, markets, and distributes the most advanced thinking on the conservation of our natural resources—books about soil, land, water, forests, wildlife, and hazardous and toxic wastes. These books are practical tools used by public officials, business and industry leaders, natural resource managers, and concerned citizens working to solve both local and global resource problems.

Founded in 1978, Island Press reorganized in 1984 to meet the increasing demand for substantive books on all resource-related issues. Island Press publishes and distributes under its own imprint and offers these services to other nonprofit organizations.

Support for Island Press is provided by Geraldine R. Dodge Foundation, The Energy Foundation, The Charles Engelhard Foundation, The Ford Foundation, Glen Eagles Foundation, The George Gund Foundation, William and Flora Hewlett Foundation, The John D. and Catherine T. MacArthur Foundation, The Andrew W. Mellon Foundation, The Joyce Mertz-Gilmore Foundation, The New-Land Foundation, The J. N. Pew, Jr. Charitable Trust, Alida Rockefeller, The Rockefeller Brothers Fund, The Rockefeller Foundation, The Tides Foundation, and individual donors.

Ghost Bears

GHOST BEARS

Exploring the Biodiversity Crisis

R. Edward Grumbine

Foreword by Michael E. Soulé

ISLAND PRESS

Washington, D.C. ❑ *Covelo, California*

© 1992 R. Edward Grumbine

All rights reserved. No part of this book may be reproduced in any form or by any means without permission in writing from the publisher: Island Press, Suite 300, 1718 Connecticut Avenue, NW, Washington, D.C. 20009.

The author is grateful for permission to include the following previously copyright material: Excerpt from "Little Songs for Gaia," from *Axehandles* by Gary Snyder. Copyright © 1983 by Gary Snyder. Published by North Point Press and reprinted by permission of Farrar, Straus and Giroux. Excerpt from "The Swan" from *House of Light* by Mary Oliver. Copyright © 1990 by Mary Oliver. Reprinted by permission of Beacon Press. Excerpt from *The Klamath Knot* by David Rains Wallace. Copyright © 1983 by David Rains Wallace. Reprinted with permission of Sierra Club Books. Excerpt from "Where All the Marmots from Marmot Pass Went To," from *Pawtracks* by Tim McNulty. Copyright © 1976 by Tim McNulty. Used by permission of Copper Canyon Press, P.O. Box 271, Port Townsend, WA 98388. Excerpt from "Damage," from *What Are People For?* by Wendell Berry. Copyright © 1990 by Wendell Berry. Published by North Point Press and reprinted by permission of Farrar, Straus and Giroux. Figure 2 is from Salwasser, Hal, et al. 1987. "The Role of Interagency Cooperation in Managing for Viable Populations" in M. Soulé, ed., *Viable Populations for Conservation*. Copyright © 1987. Published by Cambridge University Press. Reprinted by permission of the publisher. Figure 4 is from United States Department of Interior, 1992, *Grizzly Bear Recover Plan*. U.S. Fish and Wildlife Service, Washington, D.C. Figure 6 is from Noss, R. F., 1987. "Protecting Natural Areas in Fragmented Landscapes," *Natural Areas Journal* 1987, 7(1):6. Reprinted with permission of the Natural Areas Association.

LIBRARY OF CONGRESS CATALOGING-IN-PUBLICATION DATA
Grumbine, R. Edward.
Ghost bears : exploring the biodiversity crisis / R. Edward Grumbine ; foreword by Michael Soulé.
p. cm.
Includes bibliographical references and index.
ISBN 1-55963-152-X—ISBN 1-55963-151-1(pbk.)
1. Biological diversity conservation—Philosophy. 2. Human ecology—Moral and ethical aspects. I. Title.
QH75.G75 1992
333.95—dc20 92-14404
CIP

Printed on recycled, acid-free paper

Manufactured in the United States of America
10 9 8 7 6 5 4 3 2

To the Greater North Cascades ecosystem, which has helped to teach me the difference between what has been tamed and what remains wild, and to the people who are working in support of greater ecosystems everywhere.

CONTENTS

FOREWORD

Hope for biodiversity rests on the blip theory of human population. As the number of Americans inches toward a half-billion by 2150, and as our species balloons by a billion every eight years, it becomes more difficult to imagine a quick fix for the extinction crisis. Environmentalists have abandoned the goal of stopping human growth quickly; now they pray that the breeding binge will be a transient blip rather than a surge toward permanent planetary obesity. If the blip theory is correct, then there is room for optimism.

R. Edward Grumbine belongs to a long, distinguished line of naturalists who, while acknowledging the deepening crisis of biodiversity, refuse to relinquish the possibility of an ecologically wholesome future. Grumbine's scenario assumes that society will back away from the final plunge into an extinction spasm of gigantic proportions. Instead, he sets us on a long but optimistic course toward ample wilderness, and to an alternative model of civilization where some people—accompanied by wolves, bears, old trees, eagles, salmon, and herds of buffalo—live lightly in reanimated forests and plains.

The foundation for Ed Grumbine's truce with nature is built on three pillars, each of which is a mini-revolution. The first is the acceptance at all levels of policy formation of the synthetic science of conservation biology—a friendly, mission-oriented science that justifies the necessity for large areas of interconnected wildlands. The second pillar is a social/political revolution: participatory land-use decision-making at community and regional levels. The third pillar is moral, a revolution in personal and institutional actions manifesting the acceptance of a deep, biocentric ethic.

In addition to these three premises and their implications, there are many other profound thoughts in this book, and all of them tempt the mind to speculative diversions and invite the hand to marginal scribbles. But I have chosen the three pillars to help introduce this book because I agree with Ed Grumbine about their centrality.

Let us assume that all three of these mini-revolutions shall come to pass in the United States and Canada. Will these revolutions herald a new age? And if the new Eden arrives in North America, will utopia come in time to save the floras and faunas of Africa, Asia, and Latin America, where most biodiversity exists and where species are threatened with massive extinction by 2050?

These questions beg another one: how long until Utopia, both in the United States and the world? Will it take years, decades, or centuries? The answer, as always, is "it depends." In the United States, a change in attitudes along the lines of Grumbine's model could come about in a few decades. This is not to say, though, that the continent could be transformed into an Arcadian Camelot in so short an interval. Such a transformation will require restoration on a huge scale and will take much longer, at least a century.

An instantaneous ecological cure in the United States is impossible because the landscape is disrupted and fragmented. It will take some time to reverse. Cattle and sheep grazing in many of the federal lands would have to be reduced or curtailed. Road building in major sections of national forests and Bureau of Land Management areas would have to cease, and many existing roads would have to be closed. Some croplands used to produce livestock feed (worldwide, livestock consume about 40 percent of grain production) or that are now in land banks would have to be restored to nature.

A century or more may be required to reconnect wildland fragments, restoring a network of conservation corridors. Only then can grizzlies, wolves, and mountain lions reoccupy much of their former habitats. A dream, yes, but not impossible given the fullness of time.

Now is the time start. The question is, can this be done during hard economic times and when the human population of many regions of the country is growing at rates rivaling those in tropical nations? Can wilderness recovery and biodiversity protection be achieved without raiding the public treasury and making enemies of the influential real estate and extraction industries? The answer is yes, although it will be difficult and more expensive in already overpopulated areas like Florida and coastal Southern California. There are two tools for this historic project: land-use planning and a radical transformation in what most people think to be an appropriate time scale for biotic restoration.

Land-use planning has to occur at local and regional levels, not at the national level. The restoration and development of a nation-wide wildlands network depend on people with intimate experience of mountains, canyons, forests, coves, rivers, and creeks. Such planning will not work, as Ed Grumbine so convincingly illustrates, without grass-roots education and participation. Over time, each local planning group will develop a map-based program for their bioregion. Later, representatives of the regional groups will meet and integrate their local plans into a national one.

But who will pay for the forfeited profits from grazing, logging, and other uses of river, meadow, and woods? The answer lies in the second tool. In most cases no one will have to pay if our time scale is the correct one.

Say that you own a cattle ranch that sits astride a valley which forms a natural link between two massifs in Montana, and that the ranch is needed for a vital connection in the wildlands system. And say that I, a stranger, show up one day and ask you to donate your ranch to a conservancy or to the state. At first you would probably think me a fool. At worst you would feel threatened, especially if I said that a bill would be introduced into the legislature to declare your land as critical habitat, thus justifying condemnation and compensation at the fair market value. This approach wouldn't work.

But let's try another. Say that you had been a participant in a planning process, a long and sometimes difficult dialogue, just as Ed Grumbine advocates. Then you might be more disposed to the conservation objectives of the program, if not the means of realizing it. And it might make a big difference if I told you that I was not asking you to give up ranching on your land, but to consider leaving it to a conservancy after the death of your children, by which time cattle ranching would probably not be economic because the federal subsidies for grazing on public lands would have vanished long ago. Besides, there might be tax benefits to you and your children.

Or say you were the supervisor of a national forest in North Carolina, and that a section of the forest that you planned to clear-cut, against the recommendations of local conservationists, was an ideal site for a roadless corridor needed to facilitate the movement of black bear and panthers between a national park and a wilderness area. And say that I approached you and the local conservationists with the following argument. First, I would admit that the corridor did not have to be covered with old-growth forest—a second growth forest with some remnant patches of old-growth would function perfectly well, as long as it was roadless. Second, I would try to convince the local conservationists that the loss of some old growth in the short run was worth the long-term vesting of the section to the regional system of wildlife recovery. Whether I quickly could convince all parties of the wisdom of the transaction might be questioned, but time is on my side. I can wait because the biologists tell me that the bears and panthers can survive without the corridor for fifty years or so.

What about Grumbine's third pillar, the need for a behavioral revolution that would transform our species from the religion of anthropocentrism to the new belief in biocentrism? Will some kind of universal conversion experience be necessary? I think not. The conversion has already happened. The human love for plants, pets, and parks supports the hypothesis that we all have the capacity for biophilia in the same way that we have the capacity to risk our lives for others. Why, then, you should ask, is this love of nature so rarely acted on, especially at the polls?

Perhaps biocentric behavior is expressed only when one is free from fear. Maybe our task as conservationists is to remove fear, the fear in people who are threatened by attacks on their occupations, their livelihood, their world view, and their property. If fear is the problem, then we need only cultivate patience. The planning and organization, however, cannot wait.

What about the rest of the world? In Asia, Africa, and Central and South America, the momentum of the population explosion will propel human numbers higher, and more time will pass before numbers return to levels compatible with ecological and biotic sustainability. In addition, the economic difficulties of tropical societies are often deeper and more recalcitrant than ours—in part due to social inequities, the burden of debt, and the potential for violent political unrest. This leaves little hope for quick fixes. A catastrophic loss of wildlands is inevitable in the tropics—perhaps more than 97 percent of the habitat and millions of species will perish, especially if we rely only on the strategy of nature reserves.

For the tropics, the appropriate time scale for repair is much longer than in the United States. The time scale is centuries. A strategy based only on ten-year plans and thirty-year projects is a romantic delusion. To pretend that most of the national parks in the tropics will persist intact, and that their biodiversity will survive without major losses through episodes of chaos, war, and famine, is also foolish.

Let's say, for the sake of argument, that 300 years will be required to stabilize and then improve the demographic, medical, economic, political, and biotic conditions in places like Brazil, Zaire, Sri Lanka, and the Philippines. Assume a best-case scenario: By 2300 human densities will have returned to mid-twentieth century levels, to half of today's bloated 5.6 billion; by then, universal education and economic wellbeing will be entrenched and the era of political instability will have ended in the tropics. In 200 to 300 years, however, it will be too late to save more than 1 or 2 percent of the wetlands, forests, and savannahs of these countries from conversion to human use. The conversion will have happened long before, back in the early part of the twenty-first century.

Recovery of the tropics will also take longer because these nations must start to restore nature from a deeper level of destruction. Conservationists must begin now to plan for this recovery by taking the appropriate management and technological steps to protect samples of the tropical biota during the centuries-long demographic and political winter. This means we must aggressively collect, store, breed, and preserve comprehensive samples of tropical biodiversity. It means we must support research in new methods of cryopreservation and regeneration of life forms.

Some purists find such technological talk repugnant, and think that extinction is preferable to an existence without dignity in test tubes, zoos, and botanic gardens. I do not agree. As the only species capable of it, we are obligated to maintain the evolutionary process in as many kinds of plants and animals as possible.

Among the many notions that we cling to that prevent us from fulfilling this duty is the belief that biotic systems are balanced and relatively selfcorrecting: that nature, left alone, can recover from anything.

Nothing could be more misleading, especially as the planet enters an era of unpredictable environmental change. Environmental change is nothing new, of course, but the major ecological reshufflings of the Pleistocene, for example, will be surpassed in the tropics by anthropogenic loss of habitat and wildlife. Everywhere, greenhouse induced climate change, ozone depletion, and an unprecedented biotic transport of thousands of alien species and diseases to new continents will radically mix native species with introduced floras and faunas.

As a result, we will face management problems on a qualitatively different scale than in the past. Ecologists cannot predict which genes, which populations, and which combinations of species will fit the future, re-combined biotic systems. Therefore, we must reject the idea that some species are destined for extinction anyway, and we should be committed to heroic interventions to ensure continuation of the evolutionary destiny of biodiversity.

We embark on a great expedition—a Millennium Ark. Ed Grumbine is one of our navigators.

Michael E. Soulé
Santa Cruz, California

PREFACE

Students of critical thinking, whether they are active on a high-school debate team or engaged in university-level discourse, are often cautioned to be wary of inductive logic. Investigating a single part of any system and drawing conclusions about the whole can be misleading—the fable of the blind men and the elephant is an elegant reminder of the perils of this kind of reasoning.

In this book, however, the story of the plight of the Greater North Cascades stands for regional ecosystems of many kinds throughout the United States and the world. What is happening today to native diversity in one place on the planet is symptomatic of the decline of biological diversity worldwide. From one ecosystem to the next, different plants and animals are threatened, the pace of ecosystem deterioration and species extinction differs, and human efforts to reverse these trends vary, but overall the results of the biodiversity crisis are the same. An ecologist can look at the state of almost any global ecosystem with an eye toward past conditions and present trends and predict a world increasingly difficult for many life forms to inhabit.

Studying nature from an ecological perspective is provocative and always a good place to begin any story about how people work with nature. But the science of interrelationship between organisms and their environment cannot fully explain the connection between the biodiversity crisis and land management in the United States. Though I have grounded my argument for protecting biodiversity in biology and ecology, I have also attempted to provide a wide-ranging perspective across many fields of inquiry: environmental law, history, policy, management, and ethics. I was fortunate as a student to have teachers who were never much concerned with the boundaries of academic disciplines. Instead, they used knowledge as a means of gaining wholistic understanding and I trust that this book will help readers weave complex ideas into a more meaningful pattern.

Though the Greater North Cascades is the specific place from which I tell my story, I have provided examples from many other regions to highlight trends and connections that extend far beyond events in the Pacific Northwest. My wish is to reach concerned laypeople across the nation as well as environmental professionals. The task of turning technical material from numerous fields into a story that speaks to several audiences was not easy; I opted to speak from personal experience, drawing on my background in ecosystems and their management, in a style more accessible than dry academic prose.

The development of ecosystem management for native diversity will not be simple, not even in the United States, a country rich in diverse ecosystems, with a relatively low human population, a wealthy economy, and farsighted environmental laws. I am aware of the enormous restructuring of the status quo necessary if we are to make the transition to a sustainable society. I have provided few details on this subject only because I do not know how or when restructuring will occur. All I know is that it is essential for the continuation of life on Earth for many species, and that it must be accomplished sooner rather than later—much sooner.

What gives me hope is not the expectation that a more scientifically "correct" ecosystem management will soon replace our outmoded land-management practices. Rather, it is the neotropical warblers that still make their long journeys every year; the fragments of ancient forest that may yet serve as models to heal lands damaged by too much logging and road building; the Forest Service workers I've met who, in spite of heavy political pressure, continue to press their agency for change; the citizen advocates I've worked with across the country who are saying no to more development of wildlands and are working hard to ensure that environmental laws are properly implemented; the North Cascades grizzlies and gray wolves that still hang on and may someday recolonize lost habitat; and the students in my classes who are serious about ecology and working for a world healthier than the one they inherited. These are some of the many reasons to look toward the future with hope.

If this book is of any help to those who are already learning how to share Earth with grizzly bears and spotted owls, it will have been successful. If those land-management professionals working to build a better future for native species and ecosystems are encouraged to speak out more strongly for wild nature and a sustainable human way of life, my message will have found a home.

In writing this book, I owe much to the generosity of teachers and friends. Of these, none is more important than the four places that helped me "fall in love outward" at different stages of my life: the coastal beaches and forests of Baranof Island, southeast Alaska, my playground as a young child; the humid *Liriodendron* woods of the Patapsco River watershed on the western edge of the Chesapeake Bay, where I learned how to be still in the woods and how forest succession and low-impact human use may, in time, lead to thriving second growth; the Great Smokies of the southern Appalachian highlands, where I first encountered ancient forests and deep mountain wilderness; and the bright salmon rivers and dark hills of the Pacific Northwest, especially the North Cascade and Olympic mountains, to which this book is dedicated.

Many people have provided encouragement and shown enthusiasm throughout this project, and I thank them all. I could not have completed the task without so much valuable support. I wish to extend special gratitude to my undergraduate mentor Bob Bieri, who, while alive, shared his unbridled love for natural ecosystems and field ecology; Tom Fleischner, Jeff Hardesty, Tim Jordan, and Saul Weisberg, companions who walked many trails and inspired insights into friendship, natural science, and teaching; Mitch Friedman, who provided the strength to both dream and act on a vision for the Greater North Cascades; Gary Snyder, for a model of living-in-place; Bonnie Phillips-Howard, who continues to inspire me in her work as an advocate for ancient-forest ecosystems; Michael Soulé, for the Society for Conservation Biology and provocative conversations; Bob Keiter, who shared his legal wisdom; John Earnst, superintendent of North Cascades National Park, who sponsored an interagency conservation-biology workshop I presented in 1987; and Jon Almack, for his support of the North Cascades grizzly bear.

As this book is a reworking of my Ph.D. dissertation with the Union Institute, I owe a great debt to the members of my doctoral committee. John Tallmadge provided the strongest support a doctoral candidate could wish for—passion for the work, respect for the learning process, and critical expertise as overall editor and guide. Reed Noss gave generously of his time on matters ranging from conservation biology to ecosystem management to environmental values. His energy and commitment to wild nature are boundless and inspirational. Bill Devall, long-time friend and supporter, prodded me always to remember that Homo sapiens is but one of life's diverse forms. Deborah Bowman provided emotional support throughout my program. John Miles, long-time resident of the Greater North Cascades ecosystem, helped to ground my more idealistic arguments in the soil of old-growth politics.

The process of turning a dissertation into a book can be daunting. Barbara Dean and Barbara Youngblood at Island Press guided me through the final editorial stages with an eye for clarity of argument and a love of language that improved the book greatly. I also benefited immeasurably from the writing skills of copy-editor Constance Buchanan and Colleen O'Driscoll, who word-processed the many drafts of the manuscript.

Finally, I want to acknowledge my parents, whose encouragement of independent thinking and whose concern for social justice contributed to my basic beliefs, and my partner Marcy Reynolds, who continues to spark my awareness that, though the future of biological diversity appears uncertain, it is only through human acts of love and commitment that we will be able to fit into ecosystems as "plain members and citizens" and cultivate a long-term ecosystem management of place and spirit.

R. Edward Grumbine
Rattlesnake Gulch
Bonny Doon, California
April 1992

Ghost Bears

CHAPTER ONE

BOUNDARY MARKING: AN INTRODUCTION

the path to heaven
doesn't lie down in flat miles
It's in the imagination
with which you perceive
this world
and the gestures
with which you honor it.

MARY OLIVER,
"The Swan"

The center of a summer snowstorm is an unusual place to find yourself teaching a college class called Introduction to Ecosystem Management. It is even more uncommon to have a diminutive bird reveal the life in an unpeopled borderland, an international boundary closed to legal passage. But the alp land of Armstrong Mountain is no common place, the Greater North Cascades is a landscape full of lessons, and a teacher must be ready to learn.

Since 1979 I have been teaching outdoor field courses through the Sierra Institute, a program of the University of California Extension,

Santa Cruz. These academic excursions, up to eight weeks in length, allow a group of twelve students to encounter the source of "resources"—the mountains, rivers, and canyons of the West. For the last five summers I have focused my teaching on the Pacific Northwest, the region where I was born and whose ancient forests today bear stark witness to both the pain and the promise of the biodiversity crisis. For students, the lessons are powerful and clearly drawn. Here, in the North Cascades, the line between the land management of the past and that of the future is as sharp as a clearcut swath hard against a national park.

The northeastern corner of the North Cascades, before the mountains meet the lava uplands of the Columbia River Plateau, is a tumbled expanse of rolling highlands that reach north into Canada. To the east, forests of lodgepole pine crown ridges rising to timberline. Westward, the ice peaks and stratovolcanoes of the Cascade Crest stretch beyond sight. This is the edge of the mountain world and the beginning of the Pasayten Wilderness, a good place to bring students to backpack, establish a base camp classroom for a week, and learn about conservation biology and the politics of land management.

On the fourth day of one particular trip we decided to day hike above timberline to the border several miles away. The weather was unsettled, the sky alternately blue and gray, the kind that climbers eye with skepticism. But we were only out for the day, close to camp, and determined to find the posts marking the forty-ninth parallel that Canadian and U.S. Army surveyors erected years ago.

Armstrong Mountain curled away from our base camp in upper Horseshoe Basin like the shoulder of a sleeping giant, a broad hump of rock and tundra knit with grass, patches of snow, and lichens. We hiked up the swale of Snehumption Gap, the meadowy bowl that lies between Armstrong and its neighbor, Arnold Peak. The class was in high spirits. At the pass we found evidence of Pleistocene ice—the north face dropped sheer to a basin far below. Couloirs were stuffed with snow. A marmot waddled to an outcrop, stood up, sniffed the wind. We sat down on gravel-streaked ground amidst hundreds of blooming *Dryas*, the mountain avens, and faced what the maps said was Canada. There was no sign of another country; the mountains held no clue. If the students had been birds they would have soared. There was freedom in the altitude, lift to the steady wind, and a freshness infusing the sunlight sweeping across the pass. We settled in.

"What's a boundary all about?" I asked. "What does it mean to you?"

The students were quiet. The group was a good one—eager to learn, reflective, yet full of adventure—and diverse: Scott hailed from the Midwest and the University of Illinois; Lisa was a biology major from the University of California, San Diego; Kari was studying at Stanford after growing up in Texas; John, a native Californian, was completing a political science major at the University of California, Santa Cruz. All told, there were thirteen of us coming from five states and seven schools.

John broke the silence. "I think of my girlfriend. Where we meet it's like two worlds coming together. Most days it's great, but we argue too."

"Up here I can't think of anything separating things," said Kari. "I mean, look—there's no trail, no people, just mountains. Where's the border?"

"Yeah, how could they even make a border up here?" Lisa broke in. "There's nothing to mark, no trees, nobody around. Why bother? A border's got to be identified. It's like when you're inside a house and then you go outside. You put on a coat."

"So what boundaries are difficult to cross?" I asked the group. "What's been hard for you?" A big patch of blue opened in the sky to the north. The sun warmed our cheeks, the rocks, the cushion plants around us.

David, a first-year student at the University of California, Santa Cruz, looked up. "When my grandmother died I had a hard time accepting it. She brought me up and then she was gone. It bothered me a long time. I couldn't understand it."

"How old were you?" asked Lisa.

"This was just last spring, freshman year in school." David smiled. "But that seems so irrelevant here. I mean, *look* where we are!" A dark cloud covered the sun. The sky was the color of metal.

"Yeah, this is great," echoed Scott. "I can think of millions of boundaries but I don't see any right here. Besides, they help me get it together, like Lisa was saying about putting on a coat. This is a great place. What I don't understand is the difference between managing a wilderness here in the United States and not even having wilderness over there in British Columbia—like you were talking about yesterday, Ed."

"What I meant," I said, "was that B.C. doesn't have a wilderness act. That mountain you see in Canada may be undeveloped now, but it is not protected for the future. Where we're sitting is legal, capital W Wilderness, but a quarter mile away anything goes. If the two countries want to work together, right away they have a problem."

Scott stared north. "So what's the big deal? From here it still all looks the same to me. Let's go up to the top and check it out."

We packed up and hiked out of the pass. Low gray clouds were scudding in, but we were ready with sweaters and raingear.

You could almost climb Armstrong with your eyes closed. The basalt mountain is deepset in turf and glaciers have smoothed its edges to gentle humps. After a twenty-minute climb we crested onto a sea of boulders, rock polygons, and patches of tundra.

"Look, is that it?" cried Lisa. She pointed at an upright object a quarter mile away.

We walked toward the border and it began to snow. The wind gusted sleet across the flat-topped summit. When we reached Monument 104 the sun was out again, melting the squall away. The border marker was a four-foot obelisk cast of heavy aluminum. It was pitted by life above the treeline and imprinted on each side with the names of the two nations at whose common border it stood sentinel.

I looked further on and saw the gash. East of Armstrong and Arnold, where the border ran down into forest, a sixty-foot swath cleared of all vegetation demarcated the boundary. It looked as if the skin of the earth had shrunk, severing the forest. The gash climbed the ridge and plunged down the far side out of sight, reappearing straight as geometry on the next slope before it was lost in the swirling clouds of an oncoming storm. The students said nothing. Their eyes followed the truncated margin out into the mist.

It began to snow hard. We scurried west across the tundra, keeping watch for another marker on the far side of the mountain. Anger welled within me. I thought of the herbicides used, all the misplaced energy spent carving an international boundary into one of the wildest places in the Greater North Cascades. Would a grizzly cross this void? A wolf? Bears and wolves might not be stopped by the gash, but they would certainly recognize this: In the heart of the backcountry, straight as a superhighway, humans had imposed a border domestic as any lawn.

The storm raged around us as we pressed forward. It was a kaleidoscope of gusting wind, snow squalls, and shafts of sunlight. Hail pelted down, tiny flowers frosted white. The orange lichens coating the rocks glowed like the eyes of a cat. As we reached the far side of the mountain, the front of the group was bathed in sunlight while the stragglers were lost in clouds.

Monument 103 crowned the international boundary on the western edge of Armstrong. Boulders were heaped around its base. A hole in the clouds revealed the gash, slicing west, dark as any wound, rising and falling over ridge after ridge. The students were giddy and sober at the same time. We were caught between the world as it was and as it was coming to be, and everything demanded attention: miles of jumbled mountains, whirling snow, slippery footing, freezing fingers, colorful lichens, and the gash.

"I want to save all of this," Kari spoke quietly.

No one moved. The wind ripped the clouds and shut them tight again.

Then it was time to go. We left the border post, lost sight of the gash, and turned south toward camp. Suddenly the world was full of wing beats. A white-tailed ptarmigan blossomed from rocks and flying snow and settled out of sight, one bird in a land of wind and weather. We too were startled. We had been trying to escape, seeking salvation from the dark border and the freezing wind, when the ptarmigan appeared and drew us together like a magnet.

"How can it live up here?"

"Where does it go?"

"This is July. What about winter?"

Our conversation took a new course as I tried to explain what I knew about how the bird lived successfully at high altitude. We had not seen the ptarmigan at first because of its summer camouflage, what ornithologists call cryptic plumage, where a bird's feathers match the colors of its habitat. Once flushed, the ptarmigan had disappeared, feigning injury. We had scared it off a nest, and assuming we were predators, it had attempted to lead us from its young. There were other details to explain—diet, altitudinal migration, the borders of life and death for alpine birds. But by then it was too cold to lecture.

Recovered, I realized that, like the ptarmigan, paying full attention is the appropriate response to life in an uncertain world of snowstorms, fragile homes, and predators. The answer to the dilemma of humans, birds, and boundaries is not "saving" but rather healing, making whole. When people see what ptarmigans have to teach about attention, they will learn this. They may come to understand that listening is a gesture full of grace.

The wind pushed us down the mountain. I could tell from the students' demeanor that they had more questions than answers. As they picked their way down slopes strewn with boulders left by glaciers from some distant time, they stepped with care. We scared up no more ptarmigans. And, looking back, we saw nothing to preserve, only opaque clouds filming the world with mist. Ahead there was camp, the hike tomorrow, weeks of teaching, other borders to experience, questions to explore. The ptarmigan, the boundary post, the border scar receded even as the conversation they sparked among us carried on.

This is what we learned.

Miles from the nearest road, the boundary post was unexpected, out of place. It was stuck in the ground by the side of the trail in a thicket of avalanche brush: National Forest. On the far side of the post were the words National Park. I stopped in the path, searching for other signs. The salmonberries were finished for the year. Mountain-ash and elder shrubs, scattered upslope in the rubble, hung with red fruit. It was good bear habitat, although there were no tracks and no purple-tinged scat. A marmot stood by its burrow, whistling down the scree.

The boundary marker was a lone outpost in this headwater basin high in the North Cascades. Standing amidst the work of snowstorms, cloud torrents, marmots, and willows, it clearly stamped people and politics onto the wild fabric of the mountains. The line it marked ran straight across the basin with no regard to watershed. I wondered what political deals were cut in 1968 when this boundary was bargained into existence with the creation of North Cascades National Park.

A glance at a map elevates such questions to more intractable levels. First, you notice the international border, straight as an arrow, that segregates Canada from the United States along the forty-ninth parallel. The park is surrounded by a confusing mix of two national recreation areas, five wildernesses, three national forests, and a provincial recreational area, along with state and private lands. Nowhere do these boundaries reflect what happens ecologically. Neither the grizzly bear, the northern spotted owl, nor the rest of the nonhuman world recognize them.

Just as we circumscribe the backcountry, so we define and delimit our lives with borders. Private ownership implies that what is yours and what is mine can be easily distinguished one from the other. Our society stretches across a diverse continent only to order nature into a simple taxonomy: "federal," "state," and "private" lands. At deeper levels, we have come to believe in the perennial opposition between mind and body, self and other, culture and nature. Yet our uncritical acceptance of borders limits our ability to reconnect forests with parks and people with nature.

My point is not that boundaries are unnecessary. In fact, we cannot live without them. They represent visions of the world, serve as a litmus for comparing what people think nature does with what nature really is like. Problems like species extinction and habitat destruction result when human images of nature distort the patterns and processes of natural ecosystems.

This is a story of how humans can move away from an adversarial relationship with nature by understanding biological diversity and learning to live within ecological limits. Many of the details are drawn from the place that I know best: the Greater North Cascades ecosystem in Washington State (see figure 1). There is no way to learn about genes, grizzlies, and greater ecosystems without knowing a specific region, yet people intimately familiar with other places will recognize much of this story as part of their own.

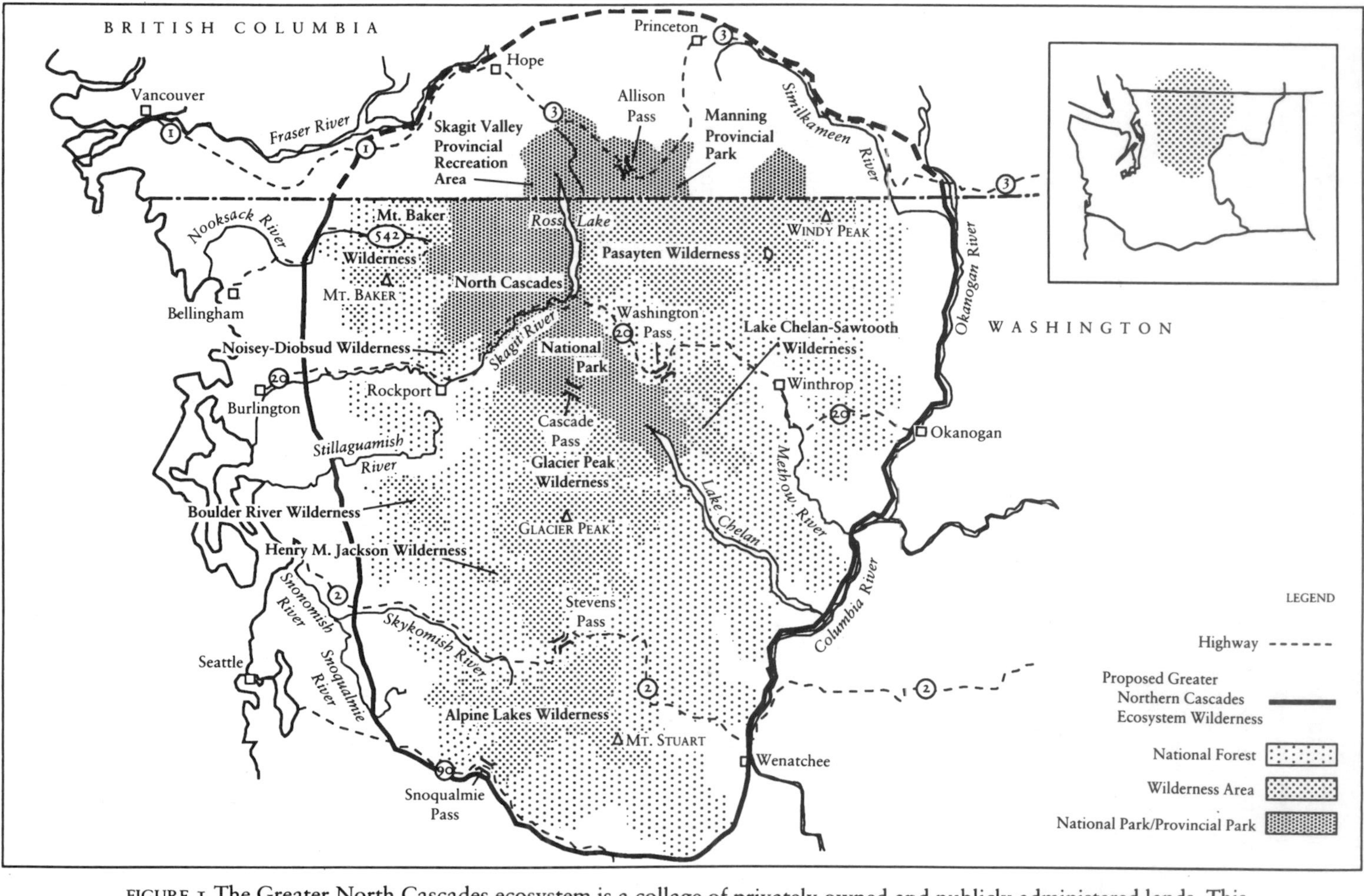

FIGURE 1 The Greater North Cascades ecosystem is a collage of privately owned and publicly administered lands. This matrix of jurisdiction represents the greatest impediment to ecosystem management.

Because most of the Greater North Cascades are administered by the U.S. government, I have chosen public land management as an entry to the larger cultural matter of how humans relate to nature. Much of the country's remaining wild habitat is on federal land; it seems obvious that if we cannot protect biodiversity there, on national parks, forests, and wildlife refuges, we will not be able to protect it on state and private lands. At the same time, as we shall see, public lands are insufficient to protect nature over the long term. Ecosystems under other ownerships will have to do their share.

The dilemma posed by the relationship between public and private lands is the beginning of the story of how boundaries drawn in the past are no longer suitable today. Misplaced borders may be ecological, legal, or managerial, but they all result from the dominant Western world view that draws a hard line between people and nature. Anthropocentricity (literally, "man centered") places people above grizzlies, greater ecosystems, and the rest of the natural world and reduces nonhumans to the status of resources for Homo sapiens. Such "resourcism" has a simple credo: The world gains value only as humans transform it into goods and services to meet their demands. We employ economics to measure such transformations and call the results progress. Three assumptions bolster resourcism: Human demands need only be met in the short term, Earth's abundance is inexhaustible, and technological savvy will continue to push back limits to growth. These attitudes color our personal interactions with nature and underlie the public particulars of federal land management.

But in the Greater North Cascades and elsewhere, resourcism is being challenged by the biodiversity crisis, the developing science of conservation biology, and a new image of working with nature called ecosystem management. The biodiversity crisis can be measured by such outward threats as the rate and scale of species extinction, loss of wild habitat to human population growth and resource consumption, increase in atmospheric carbon, and ozone depletion. Meanwhile, these threats are creating a psychological atmosphere of tension in humans that may eventually help people to reevaluate how they relate to nature. The biodiversity crisis may spur a revolution in human attitudes toward the natural world.

With extinctions increasing and habitats being destroyed, biologists have begun to explore the implications of the loss of biodiversity through conservation biology. Conservation biologists seek to understand the dynamics of extinction, the specific factors that place organisms from bears to salamanders at risk. And, because all populations require some habitat to survive, biologists are attempting to portray what a long-term, fully functioning system of nature reserves would look like. They are reevaluating the concept of the balance of nature from the cellular to the biospheric level and assessing the implications of managing nature from the perspective of resourcism. The new science reveals how radically different management must become if we expect to continue to inhabit a functioning planet. These issues provide a scientific argument against resourcism and for a better working relationship with the Earth.

The grizzly bear is the totem of the biodiversity crisis in the Greater North Cascades. As the largest and most powerful mammal in the region for the last ten thousand years, the bear has been, until recently, a respected teacher of Homo sapiens. The story of the grizzly provides compelling evidence of how resourcism has damaged the relationship between humans and one species in particular. Endangered bears and information drawn from conservation biology undermine the prevailing view that the safety net of U.S. environmental laws is adequate to protect biodiversity. Later we will explore the legislation behind land management from the perspective of grizzly bears, spotted owls, and biodiversity at many levels. Current laws are deeply flawed, and legal reform is critical if we are to make our relations with nature more sustainable.

It is difficult for land managers charged with implementing environmental laws to deny the emergence of conservation biology, the precarious state of the grizzly, and the inadequacy of current laws. But resourcism still has a grip on most agency professionals; the force of new facts does not automatically enlighten the land-management bureaucracy. The biodiversity crisis, so far, has spawned a narrowly conceived form of scientific ecosystem management that continues to advocate value-free science, control by professional experts, and centralized decision making with little input from citizens. Toward the end of this book I will offer a specific platform for ecosystem management for native diversity, connecting conservation biology with legal, managerial, and community-based participatory options for long-term change. These recommendations are my attempt at integrating a scientific, bottom-line answer to the biodiversity crisis with an ethical alternative to resourcism, an image of nature where humans may learn to value greater ecosystems as much as they value themselves.

Mexican activist Gustavo Esteva suggests that the biodiversity crisis may finally result in humans replacing the myth of progress through exploiting natural resources. Implementing ecosystem management for native diversity will not be easy, but the biodiversity crisis has reached such proportions that there appears to be no other way to sustain the ecosystems upon which all living beings depend. The biodiversity crisis becomes the ultimate test of whether people will learn to fit in with nature, and ecosystem management gains importance far beyond the issue of finding new ways to manage national parks and forests.

The promise of the biodiversity crisis is this: Adjusting land management to stave off mass extinctions and global habitat destruction will not only help us to reduce our impact on the biosphere but also allow us to reinterpret our place on the planet as one species among many. It is possible for an ecocentric image of nature to evolve out of an ecosystems view of land management. This may take a long time, but it is being fueled today by the fate of species like the grizzly bear and places like the Greater North Cascades.

Human attitudes toward nature cannot escape modification by the biodiversity crisis. This is happening right now, if slowly, wherever conservation biology has caught the ear of managers and activists. Our nature-as-storehouse value system will certainly give way to a more appropriate image of the world based on the patterns and processes of ecosystems. Human societies will become more sustainable as people adjust to living within the limits of a finite planet. I do not believe, however, that an ecosystem view will be sufficient to sustain life into the future. People must learn to celebrate diversity as well as protect it—to revere the grizzly bear as a teacher and not just an endangered species. We must learn a wholistic approach: Species are not separate from their habitat; human laws must match the laws of nature; science is shot through with values; and people must fit in to places. After centuries of attempting to dominate nature in the West, we cannot assimilate these basic truths by simply invoking viable population theory, an endangered-ecosystem act, ecosystem management, or environmental ethics. I hope my book will clarify this. The biodiversity crisis not only calls on humans to integrate science, law, management, and ethics; at the deepest level, it urges us to pay attention to where we are on the trail, to listen as both teachers and students, and to foster our discovery of gestures that honor the world.

CHAPTER TWO

THE BIOLOGY OF THINKING LIKE A MOUNTAIN

As the crickets' soft autumn hum
is to us
so are we to the trees
as they are
to the rocks and hills.

GARY SNYDER,
"Little Songs for Gaia"

On any clear day from the summit crater of Mount Baker, the North Cascades range is transformed magically. The spectacular climax of old-growth forests, wild, glacier-gouged valleys, and endless unpeopled mountains that most humans know as North Cascades National Park expands into a larger world of oceanic clouds and geologic strata thrust skyward by the force of drifting continents. In fertile lowlands below, between the mountains and the sea, people cluster in cities and suburbs and spread up broad river valleys in towns and on farms. This sweeping view of a mountain range of inspiring size and wildness is experienced as but a small part of the planet—it can be teased from a map, almost touched from an airplane—but you still feel, standing on the peak, as if you are at a still point of the turning world.

The Lummi Indians knew Mount Baker as Koma Kulshan, the "Great White Watcher." What does the mountain see? Is there hidden meaning in the cloud masses that roll eastward off the Pacific and drench the western slopes of the Cascades with life-giving snow and rain? What is it like to participate in the creation of weather, to be a working part of the North Cascades? What wisdom may be gained from standing in one place for 700,000 years, experiencing time at the pace of millennia? Aldo Leopold, in *A Sand County Almanac*, believed that "only the mountain has lived long enough to listen objectively to the howl of a wolf." If this is true, humans, with their view of nature limited by a short-term, seventy-odd-year stay, could learn about the long run from Koma Kulshan. If we could somehow query the mountain about the health of the Greater North Cascades neighborhood, what would the nature of its reply be?

I imagine Koma Kulshan would provide several answers, and at different scales of space and time. To a smoking volcano, a 600-year-old Douglas-fir lives for a month and a 40,000-acre forest fire is an itch on an elbow. To facilitate communication, the mountain would have to telescope its vast experience into terms and a time frame meaningful to humans, say, the last few hundred years.

To begin, the mountain might note that only 3.2 percent of the Earth's surface has been protected in parks, reserves, and wildlife sanctuaries.[1] A little over 4 percent of the United States is protected as wilderness, leaving about 95 percent open to development. At the scale of Washington State, about 12 percent of all lands are set aside. Throughout the world, the country, and the Pacific Northwest, mountain ranges are relatively well protected. A glance at any map shows that parks and reserves are clustered in the scenic high country. Koma Kulshan is part of the Mount Baker Wilderness and the North Cascades are chock-full of parks, wildlands, and roadless areas. What does this mean for ecosystems outside of the highlands? Because of their relatively harsh climate, mountains, in general, do not harbor the diversity of life that is found in river valleys, foothills, marshes, and plains. In California, for example, 95 percent of alpine ecosystems are protected while less than 1 percent of riparian communities, which have the greatest species diversity, are in reserves.[2] This pattern, found throughout the world, has evolved because humans proclaim preserves based on how they look aesthetically, not on how they work ecologically. Adopting a mountain's perspective might help us to consider the long-term implications of this practice.

Turning from the high country to human cities in the lowlands, Koma Kulshan would ponder the population growth of Homo sapiens. Seattle and the cities of Puget Sound are booming. The growth rate of the United States as a whole is beginning to edge upward again. On a planetary scale, the United Nations predicts that thirty-five years from now humans may number 8.5 billion, with the increase since 1950 equaling the total increase over the millions of years since the species first emerged.[3] Demographers suggest that if present trends continue, the world's population will probably level off in about a hundred years at around 10 to 12 billion people, double what it is today. Some biologists have already calculated that it will take five hundred to one thousand years before human numbers drop back down to an "optimum" level, barring famines, epidemics, and wars.[4] What havoc would these numbers wreak on the rest of Earth's life forms?

Though the link between population growth and resource consumption is sometimes difficult for people to discern, it would seem clear as unsullied air to a mountain. The old-growth forests at the foot of Koma Kulshan in the western North Cascades are falling faster than they can grow back. On all the westside national forests in Oregon and Washington the cut level mandated for 1991 was 3.45 billion board feet (bbf), yet studies taking ancient-forest ecosystem protection and the northern spotted owl into account suggest that less than half that amount could be cut on a sustained yield basis.[5] At the global level, resource consumption faces obvious limits. As biologists Paul Ehrlich and E. O. Wilson make abundantly clear, if growth in human population is multiplied by per capita aspiration for material goods, the future for undeveloped lands is grim.[6] Topsoils worldwide are being eroded at many times the rate at which they are formed. The United States is losing excessive amounts of soil on 44 percent of its cropland.[7] On another front, if the people of India owned automobiles on the same per capita basis as Americans, the subcontinent would be clogged with 440 million vehicles, almost four times the number that exist in the United States today.[8] Global oil and gas consumption and atmospheric pollution would be radically increased.

Koma Kulshan's eye sweeps south of the Columbia River and north to the dry ranges of interior British Columbia, where the Cascades settle into foothill forests of ponderosa pine. It notes that the North Cascades grizzly bear population is down to maybe ten, that salmon runs are greatly diminished, that scores of Northwest species are at risk, and that throughout the United States, thousands more are on lists labeled endangered. Globally, plant and animal extinctions may run into the hundreds of thousands owing especially to the loss of tropical forests.[9] Why does it matter that grizzly bears, northern spotted owls, tailed frogs, and tropical rainforests continue to exist?

A mountain's memory is rich and long. Eleven thousand years ago, Koma Kulshan experienced North America's loss of 74 percent of its genera of large mammals.[10] Throughout the geologic record there is evidence of at least five to six megaextinction periods when numerous species went under. What makes extinction today different from these past events?

The answer lies with the rate, scale, and cause of the current loss of species. Reaching back beyond the birth of Koma Kulshan 600 million years ago, earth history reveals a "background" rate of extinction of roughly one species per year.[11] Today, that rate is hundreds, possibly thousands, of times higher. Life forms are winking out all over the planet. Past extinction spasms, triggered by climate change, occurred over long periods of time, 10,000 to 100,000 years. Human activity accounts for most extinctions today, and the speed of loss is breathtaking—15 percent of all the world's species could be gone in ten years.[12] One rough estimate suggests that the following groups will all but disappear within the next hundred years: primates (excepting humans), the large carnivores, the nonruminant ungulates (camels, rhinos), and most of the hoofed animals.[13]

No biologist can foretell the effects of this potential destruction, and Koma Kulshan's history may not go back far enough for clarity. We do know, however, that extinction does matter, and for at least two reasons. As the primary agents of species loss, humans have the power to stop most extinctions from occurring. There are ethical arguments that conclude that what we are doing is wrong (see chapter 7). The second reason may be more immediately compelling: People are utterly dependent on biological diversity. Food, water, and all the necessities of life are derived ultimately from functioning ecosystems. Lose enough species from any system and it is likely to become dysfunctional in unpredictable ways.

There is another distinction between extinctions past and present. It was only in the mid-1970s that many scientists began to worry about the wholesale loss of species. Today, anthropogenic (human-caused) damage seems to be moving to another level. Biologists are now concerned that ecosystems themselves, and not just the plants and animals that inhabit them, are endangered. Looking down across the westside Cascade forests, one can only agree. At least 90 percent of old-growth Douglas-fir forests have been logged.[14] Only about 6 percent have been protected, and there is a great dispute over the fate of that small remaining percentage. Other ecosystems are also at risk—wetlands throughout the United States have been reduced by 50 percent, about 40 percent of the small lakes in New York's Adirondack Mountains are now fishless, and most of the Midwest tallgrass prairies and Florida's longleaf pine forests are no more. And, though 85 percent of the neotropical forests of Brazilian Amazonia remain intact, at current rates of logging they too will disappear within a matter of decades.

Anthropogenic change does not stop at the level of ecosystems. Koma Kulshan's sight is increasingly obscured by air pollution drifting south from Vancouver and north from Seattle. By 1989, sulfur dioxide and ozone damage to plants and trees had been recorded in fifty U.S. national parks, and the problem is expected to worsen.[15] Ecologists are scrambling to understand how ecosystems will respond to air pollution near the surface of the planet as well as the effects of global warming on the upper atmosphere. Predictions, so far, are sobering. Even at the low end of projected worldwide temperature increases, forest species, patterns, and processes would all be affected.[16] At the scale of the biosphere, photosynthesis and carbon and water cycles will likely be altered. At the regional scale, species migration rates will be hampered. Ecologist Margaret Davis of the University of Minnesota has tracked how fast tree species migrate through seed dispersal as regional temperatures fluctuate over thousands of years.[17] She believes that Midwestern trees such as red spruce and sugar maple would probably have a difficult time keeping pace with the rapid rates of environmental change brought on by the increasing temperatures predicted over the next thirty years. The trees would have to spread ten times faster than their fastest recorded rate to adapt to changing temperatures and environmental conditions.

Fires, droughts, and other natural disturbances would likely combine with changing species migration rates to alter profoundly the composition of forest communities at the ecosystem level. And, for individual trees, time of flowering, leaf drop, seed dispersal, and growth rate would all be subject to variation due to increased atmospheric carbon.

The mountain senses continual change. It has witnessed the slow turning of temperature and climate. It remembers when Douglas-fir forests were just beginning to dominate the Pacific Northwest thousands of years ago. Koma Kulshan gazes down at a twenty-four-year-old phenomenon, North Cascades National Park, and is itself encompassed by a protective legal designation, wilderness, all of eight years old. In the context of the mountain's geologic age, the laws that regulate parks and reserves are hopelessly ephemeral. Employing a more meaningful standard, say the span of a century, to compare environmental change with human legal mandates, many scientists would agree with ecologist Walt Westman's remark that today's land-management classifications are "based on concepts that are no longer widely accepted in the ecological community."[18]

Koma Kulshan sees what biologists are calling the biodiversity crisis: inadequate nature reserves, human overpopulation and nonsustainable resource consumption, species extinction, endangered ecosystems, impending rapid climate change, and imperfect laws. But the mountain cannot communicate in human language. It can only offer itself as a silent teacher, a model of long-term residence. From the valleys of the Nooksack and Skagit rivers, the mountain is a lofty cone of perennial snow or a massive gray presence swathed in clouds. Cloud-wrapped or clear, it could serve as a model, a steady reminder that each clearcut, road, and town must be placed with care.

To think like a mountain is to perceive different levels of biodiversity. The biosphere is present in the horizon's curve and in banks of atmospheric clouds. The regional landscape is knit from tidewater to timberline and beyond. At the ecosystem level, avalanche shrubbery is distinct from montane forest, valley-floor old growth from hayfields, and interstate highway margins from cottonwood thickets. The track of a solitary grizzly bear crossing a snowfield brings news of species and populations. The chromosomes that code the genetic level are hidden in the cells of all life in the North Cascades. Koma Kulshan embraces all these levels of biodiversity at different scales of space and time. The time of hatching mayflies is meshed with the time of melting glaciers.

Little is missing from Koma Kulshan's view. Meanwhile, the human view remains incomplete. People do not literally think like mountains; we see nature narrowly, Leopold believed, because we have not learned to develop an ecological perspective to acknowledge landscape patterns, and to fit in with nature as "plain members and citizens." This myopia cannot persist. The biodiversity crisis is now forcing us to remember that we are one species among many, that species are but one of nature's levels, and that twenty years of park protection is trivial in the context of the evolution and long-term health of ecosystems.

Mountains are not mute. Koma Kulshan does not speak in a human tongue, yet it embodies a message that spans the North Cascades from the Columbia River to Puget Sound, south to the Siskiyous and north to the limits of ancient forest. With a view reaching back before the beginnings of human habitation in the Pacific Northwest, the mountain might give us this message: "Keep watch across space and time. Remember the distant future. All that fits will remain."

Defining Biodiversity

Biodiversity, according to the U.S. Office of Technology Assessment, is "the variety and variability among living organisms and the ecological complexes in which they occur."[19] There is, however, more to biodiversity than numbers of species and kinds of ecosystems. Ecologist Jerry Franklin portrays ecosystems as having three primary attributes: composition, structure, and function.[20] Ecosystem *components* are the inhabiting species in all their variety and richness. Many different species, gene-pool abundance, and unique populations are what most people think of when they hear the term *biodiversity*. But there is more to consider.

Ecosystem *structure* refers to the physical patterns of life forms from the individual physiognomy of a thick-barked Douglas-fir to the vertical layers of vegetation from delicate herbs to tree canopies within a single forest stand. An ecosystem dominated by old, tall trees has a different structure than one comprised of short, quaking aspen. And there is more structure in a multilayered forest (herbs, shrubs, young trees, canopy trees) than in a sagebrush grassland, prairie, or salt marsh. Across the Baker River watershed, groups of forest stands may even exhibit a kind of patchiness or landscape "grain": broken-topped old growth here, a uniform tree farm there, and streamside deciduous alder interspersed with open gravel bars following the river's flow.

Ecosystem *functions* are hard to see in action. "You can't hug a biogeochemical cycle," says one ecologist.[21] But without the part of the carbon cycle where small invertebrates, fungi, and microorganisms work to break down wood fiber, the downed logs in an ancient forest would never decay. Natural disturbances also play a role. Wildfires release nutrients to the soil, weed out weak trees, and reset the successional clock. The energy of falling water creates spawning beds for salmon even while it carves a mountain's bones. Plants breathe oxygen into the atmosphere. Ecological processes create landscapes and diverse environmental conditions out of life itself.

Ecosystem components, structures, and functions are all interdependent. To understand biodiversity, one has to think like a mountain and consider not only the biotic elements of plants, animals, and other living beings, but also the patterns and processes that shape volcanoes and forests.

Since Aristotle, scientists have conceived of life as a biological hierarchy. Many people (and not a few biologists) have mistaken the familiar levels of biological relationship—genes, species, ecosystems, and the biosphere—for an ethical statement, constructing a Great Chain of Being out of the differences between birds and forests, apes and humans. The biodiversity hierarchy describes the complexity of life; it does not ascribe value. We can expand our understanding of biodiversity by using hierarchy theory to illuminate the relationships among genes, grizzlies, and greater ecosystems.

The biodiversity hierarchy is composed of a genetic level, a species-population level, a community-ecosystem level, a regional or landscape level, and the biosphere. To understand these levels and how they interact, consider the grizzly bear. At the *genetic* level, the grizzly has a unique set of genes that has allowed it over evolutionary time to adapt to local environmental conditions. A North Cascades grizzly "fits" the land differently from a Yellowstone bear. A grizzly's genes provide some defense against local pests and pathogens. And they define the future by investing the individual bear with flexibility in responding to changing environmental conditions over time.

A population of ten grizzly bears pools a certain amount of genetic diversity. A population of 200 bears likely contains even more diversity and is therefore better equipped to adapt to the future. The *species-population* level of biodiversity encompasses the attributes of individuals assembled in groups, which population biologists call demes. One can follow the fate of the North Cascade grizzly deme, the Yellowstone deme, the Yukon deme, in other words, the different subpopulations of grizzlies as a whole. Each of these groups will have a certain census size, number of males and females, birth-death ratio, growth rate, and other such demographic characteristics.

Grizzly bears range over tracts of territory hundreds of square miles in size. As they travel they encounter many distinct vegetative communities—low-elevation pasture, lodgepole pine forest, mountain meadow, subalpine spruce and fir, and land above the treeline. The *community-ecosystem* level of biodiversity encompasses the different kinds of plant and animal groups found in such communities.

Ecosystems change over space and time. An old-growth Douglas-fir stand may include an acre of red-cedar swamp. Where a giant fir has fallen, a hole in the canopy lets light pour in to which quick-growing plants respond. With time, the gap will close and a small pocket of second growth will fill the space, surrounded by trees that count the years in centuries. A 40-acre clearcut creates a larger patch and a 400-acre forest fire one larger still. Where fires burn through duff to mineral soil, trees may not grow for a hundred years. But wildfires do not burn evenly over the ground; in their wake there will be patches of untouched greenery that may continue to grow as if little had occurred. With communities composed of a mosaic of such patches at different scales of space and time, it is always somewhat arbitrary to set specific ecosystem boundaries. Nevertheless, it is easy to recognize the differences between ancient forest, subalpine meadow, and sagebrush.

As wide-ranging mammals, grizzlies inhabit a regional or greater ecosystem made up of many kinds of ecosystems. At this scale, the *regional or landscape* level of biodiversity, distinctive patterns between ecosystems emerge. Each old-growth forest exists within a regional context—the links between separate stands are as important as the species "content" of any stand itself. If one patch of ancient forest harboring spotted owls is cut off by clearcuts from a neighboring stand beyond the dispersal distance of owlets, the population may be more susceptible to extinction. For bears living on the east side of the Cascades, new logging in roadless areas may sever travel corridors between subalpine winter dens and early-spring food sources in the lowlands.

The regional mosaic of ecosystems contains gradients of multicommunity diversity that have been little explored by ecologists. Every bear biologist knows that no subpopulation of grizzlies in the continental United States is by itself large enough to be considered viable. The best estimate of the North Cascades deme, for example, is ten to fifteen animals, while the nearest neighboring population in the Selkirk Mountains of Washington, Idaho, and British Columbia harbors some twenty to forty individuals. Outside of captive breeding or trucking bears from one area to another, the only solution to this particular landscape-scale dilemma, where human developments have fragmented what was once a single population, is to reestablish the biological corridors that once connected the two subpopulations.

The *biosphere* is the highest level of biodiversity. Life, as far as we can tell, is limited to Earth. Planetary patterns of diversity are evident from the equatorial belt of tropical forests to the tundra and ice encircling the poles. Emissions from the burning of fossil fuels affect this level of biodiversity as the atmosphere becomes loaded with carbon and as global temperatures rise.

Hierarchy theory suggests that what happens at higher levels of organization will constrain lower levels.[22] As ecologist Reed Noss points out, "If a big ball rolls downhill, the little balls inside it will roll downhill also."[23] Global warming threatens to alter many of the planet's ecosystems. Human-generated landscape change across the Greater North Cascades isolates grizzly demes and severs old-growth ecosystems. Species populations become subpopulations inhabiting patches of what was once continuous ancient forest. Over time, as a result, the genetic variety of a given subpopulation could be reduced.

A grizzly bear inherits a specific set of genes that confer individual traits, yet each bear is also a member of a population. Arthur Koestler, contributing to hierarchy theory, coined the "Janus principle" to describe this dualism: All parts of the biodiversity hierarchy "manifest both the independent properties of wholes and the dependent properties of parts."[24] Hierarchy theory also proposes that each level of biodiversity contains emergent properties that are unique to each level—populations have characteristics that individuals do not share, and so on up the scale. A single grizzly has a genetic identity but not a birth-death ratio. A grizzly population has a demographic ratio of births to deaths, but it does not have an index of species diversity. An old-growth forest contains numerous species but, by itself, exhibits no biological corridors across landscapes.

The biodiversity hierarchy and Jerry Franklin's tripartite conception of ecosystems may appear to offer straightforward models of nature, but several human biases work to skew our perception of the diversity of life. People tend to focus on the species-population level and discount the other levels. There exists, for example, an Endangered Species Act, but no endangered ecosystems act (see chapter 3). On the species level, we focus on individual plants and animals that grab our attention: bears, blue whales, primates, towering Douglas-firs, redwoods. These are the "charismatic" megavertebrates and floras. Humans discount species that are difficult to interact with, that do their work unseen—spiders, soil invertebrates, mycorrhizal fungi, insects, and so on. Problems arise when we expect our limited view to capture the complexity of life at all levels. A nature reserve in the United States built around a single species of migrating warbler will probably not serve the bird well over time. A species approach to conservation cannot alone protect biodiversity across the breadth of nature.

When people do attempt to grasp all the levels of biodiversity, there is a tendency to separate ecosystem components from patterns and processes, communities from landscapes. Modern humans have an easier time with parts than with wholes and often forget that, in nature, interrelationship is as important as individuality. Foresters, for instance, have sought to maximize fiber production from species like Douglas-fir and ecosystems such as ancient forests, but they have neglected the role of root fungi, bacteria, and soil structure in growing new generations of trees. Many trees allocate over 70 percent of their photosynthate to below-ground roots and mycorrhizal fungi—the maintenance of a living rhizosphere, or root community, depends on intact forest ecosystems above ground.[25]

The spatial complexities of biodiversity at many levels have not often been defined or determined. Short-term solutions in forestry disregard the soil as a critical link in the ongoing health of forests. Spatial scales are also important to remember in designing parks and preserves—a half-mile may be seventy-five times the "cruising radius" of a mole, but it represents only 1 percent of the area a solitary cougar traverses.[26] The original boundary of Yosemite National Park included both high-country summer range and low-elevation habitat for mule deer, but when Congress reduced the park's size in 1904 to spur development, the deer population lost most of its protected wintering grounds. At the landscape scale, park managers look at grizzlies and old-growth forests inside reserve boundaries and discount the ecological connections *between* protected areas. These connections may also exist across larger areas. The migrating warbler mentioned above needs not only a protected home in the United States but also refuge in the Central American forests where it lives during winter. And, if hierarchy theory is correct and there are emergent properties at each different level of biodiversity, focusing on any single level to the exclusion of others will be misleading. A Mexican ornithologist trying to protect a warbler by studying its genetic character must also remember to look at the habitat protection the bird has in the United States.

Humans also have difficulty grasping concepts of ecological time throughout the biodiversity hierarchy. Life at higher levels of biological organization, in general, cycles more slowly than at lower levels. A western hemlock produces sugars through photosynthesis on a daily basis, while the forest ecosystem of which it is a part ages over decades. In some Northwest forests, high-intensity, stand-replacement fires sweep through on an average of one every 465 years, and the regional landscape is transformed.[27] Without human disturbance, climates change over tens of thousands of years. When we constrict time to short-term periods, say, five to one hundred years, we tend to see balance in nature rather than a dynamic, changing world. Natural disturbances such as wildfires, windstorms, and floods may be catastrophic in human terms, but they are part of an ongoing, nonequilibrium flow over ecosystem time. The myth of stable environments has given us national park management based on preserving "primitive vignettes," ecosystems as living museums.[28]

What is the result of such narrow human conceptions? Ecologists Reed Noss and Larry Harris suggest four ways in which nature-protection efforts are limited. "Conservation

A) is static (it does not deal effectively with continuous biotic change);

B) focuses on individual parks and reserves (*content*) instead of whole landscapes (*context*);

C) focuses on populations and species instead of the larger systems in which they interact, and/or relatively homogeneous communities rather than on heterogeneous mosaics, and;

D) is oriented toward maintenance of high species diversity instead of characteristic "native diversity."[29]

From the standpoint of ecosystems, the biodiversity crisis exists because of the peculiar human image of nature that emerges from this composite view. Scientists have contributed their share to a species-heavy picture of life within ecosystems that are in harmonious balance. There are, however, bears, forests, and mountains that would advise us otherwise. The testimony of endangered species, fragmented forests, and smoldering volcanoes is increasingly difficult to ignore, and biologists are beginning to take heed. Now, more realistic images of nature are being formed. The biodiversity hierarchy is interdependent with ecosystem patterns and processes; maintaining life requires the continuation of underlying functions such as nutrient cycles, natural disturbances, global water and climate processes, and evolution itself. That these basic processes of the earth household are now being threatened is reason enough to pair the word *crisis* with biodiversity.

Koma Kulshan's world is our world as well. This world, we are learning, is dynamic, ever changing, and only at particular scales of space and time, in equilibrium. Much of this new image of nature results from the scientific study of the biodiversity crisis, conservation biology. When biologists try to think like mountains and look into the future, this is what they see.

The Advent of Conservation Biology

My formal introduction to conservation biology began when I attended the National Forum on Biodiversity in Washington, D.C., in 1986. But I did not start to understand the revolutionary implications of the new discipline until I had spent several years scrambling through clearcuts, arguing with Forest Service managers, mapping roadless areas, and dreaming of functional landscapes and greater ecosystems, all in a specific place, the Greater North Cascades. If you wanted to set up a sustainable system of nature reserves in the Greater North Cascades and beyond, I wondered, what would you need to protect and how would you go about doing it?

This question appears simple on the surface but has gained in complexity as the field of conservation biology has matured. What began as an attempt to integrate scientific theories of island biogeography into conservation practice has grown far beyond the original queries.

Conservation biology is the science that studies biodiversity and the dynamics of extinction. Much of this work focuses on how genes, species, ecosystems, and landscapes interact, and how human activities affect changes in ecosystem components, patterns, and processes. What species are vulnerable to anthropogenic change and why? Can we act to protect them, and if so, how? What are the implications of the biodiversity crisis for people and other life forms? These are some of the questions that fascinate and frustrate conservation biologists.

Conservation biology is an applied science. It differs from other natural-resource fields such as wildlife management, fisheries, and forestry by accenting ecology over economics. Most traditional resource management is reductionist, mainly concerned with species of direct utilitarian interest: How can humans have bucks to bag, trees to harvest, salmon to catch? Conservation biologists, in contrast, consider the entire biodiversity hierarchy at diverse scales of space and time and generally "attach less weight to aesthetics, maximum yields, and profitability, and more to the long-range viability of whole systems and species."[30]

The equilibrium theory of island biogeography proposed by Robert MacArthur and E. O. Wilson in 1967 first prompted biologists to consider population dynamics and extinction in relation to the size of nature reserves.[31] Biogeography looks at the distribution and dispersal of species, and MacArthur and Wilson, noticing certain patterns unique to islands, suggested that the number of species inhabiting a given location depended on the ratio between extinction and immigration. The two ecologists proposed that immigration rates depended on how far away an island was from a continental source pool of potential immigrant species. Extinction rates were correlated with the size, or area, of the island. Their reasoning was straightforward: All things being equal, large islands encompass a greater diversity of habitats and ecosystems, which yields more species. And, because these islands have more area, species populations on them tend to be larger. Both these factors depress extinction rates. Small islands usually have less ecosystem diversity, fewer species, and smaller populations. Extinction rates would therefore tend to be higher on small islands. Overall, MacArthur and Wilson implied that the area of any island could provide a rough estimate of how many species one might expect to find there.

Population biologists were quick to see the potential usefulness of island biogeography theory in conservation practice. As human destruction of natural ecosystems accelerated, parks and reserves were increasingly cut off from their ecological context and isolated as if they were islands in the sea. For species dependent on intact habitat, island biogeography theory might allow biologists to predict extinction rates for protected areas.[32] This was the "island dilemma": How big did a reserve need to be to maintain species diversity over time?

As the dilemma was investigated, it became apparent that simply measuring an island's area was not sufficient to explain extinction rates. First, there are several kinds of islands. Oceanic islands have always been isolated from continental land masses or source areas. The Hawaiian archipelago, born of volcanoes in the middle of the Pacific Ocean, is the classic example. Other islands, the land-bridge type, were once part of a continent but became isolated when sea levels rose, flooding connecting land. Land-bridge islands, in general, have been found to contain more species than oceanic islands of similar size. They also lose species at a faster rate after isolation. These observations have led biologists to the concept of relaxation rates. Once isolated, a land-bridge island continues to lose species until a new equilibrium is established. The elapse of time since isolation becomes an important factor in understanding extinction.

Scientists noticed other patterns of extinction as well. Not all species are created equal; some are more extinction-prone than others. Certain species tend to disappear early on—those with large bodied members, low population growth, small numbers of individuals, poor dispersal ability, and complex social structures. Biologists realized that to fathom extinction, "species must be weighed, not just counted."[33]

Islands are not simply theoretical "space." That concept was an abstraction until biologists explored the varieties of ecosystems on islands. Not surprisingly, they discovered extinction rates to be affected by the kind, quality, and diversity of available habitats. Rates of succession and natural disturbance also played a role.

As biologists compared reserves to islands, the complexity of the issue grew. Protected areas might constitute a third class of islands—habitat islands. Take Mount Rainier National Park, in the central Cascades south of Seattle. Like a land-bridge island, Mount Rainier is slowly being severed from its "continental source pool" of ancient forests by logging roads and human population growth. Since 1920, data suggests, the park has lost 26 percent of its resident mammals.[34] Is this evidence of a relaxation rate that will further reduce the park's species over time? Mount Rainier is not yet completely cut off from its regional context. It may be twenty to thirty years before it is effectively island habitat for many species. Will present rates of habitat fragmentation continue? Is the data on mammals collected in 1920 empirically sound? Is it possible to foresee a time where old growth is completely gone from all but the protected areas of the Cascades, and Mount Rainier becomes its own continental source pool?

Early work in the field has progressed, and conservation biologists have moved away from testing island biogeography theory toward two fundamental questions: How low can you go?, and How big is big enough? Geneticists and population biologists, concerned about species loss, have asked the question, What is the size of a minimum viable population? Ecologists working at the community and landscape levels and noting habitat fragmentation have asked, What size, shape, and type of connections will maintain viable populations in a system of protected areas? Viable populations and reserve size represent two paths toward the same goal, protecting biodiversity; at some point, a minimum viable population needs a certain amount of habitat to survive. Conservation biology is providing a scientific answer to my original question—What would it take to protect the diversity of life?—and a biological basis for thinking like a mountain.

The Quest for Viable Populations

Some people are fond of risk; their motto is Go for it! Others prefer to take no chances, they forgo adventure to maintain the status quo. Reason would suggest that both positions are valid, conservation biology that neither is completely so. The conditions of life on Earth make it difficult for living things *not* to go for it, and chance is impossible to avoid. As biologist Mark Shaffer observes, "The life of any individual organism is a probabilistic phenomenon."[35] While I was doing research on this book at the University of California, Santa Cruz, library, the Loma Prieta earthquake reduced downtown to rubble, an event I had obviously not expected. Uncertainty arises from random changes in natural systems. It is not, however, unrelated to deterministic processes. Freezing temperatures in autumn cause colorful leaves to fall in deciduous alder and cottonwood forests along streams in the North Cascades. Leaf drop is deterministic and predictable, guaranteed to amaze and delight us year after year. But predicting exactly where any one leaf will land is another matter. Depending on the breeze and its own buoyancy, which may have to do with how much it has been eaten by insects, the leaf may drop straight down or sail hundreds of feet away. Where a leaf settles is the result of uncertainty.

Extinction, conservations biologists have found, is both deterministic and uncertain.[36] Cause-and-effect extinctions, however, are relatively easy to predict. Remove all the sources of carrion from a turkey vulture's habitat and the bird will cease to live there. Introduce six house cats into a backyard ecosystem and avian populations will soon disappear. Stochastic (uncertain) extinctions, on the other hand, are much more complex. In an effort to understand how populations persist in the face of random environmental change, conservation biologists have grouped stochastic processes into four broad classes:

1) *Genetic uncertainty*, or random changes in genetic makeup due to inbreeding, etc., which alter the survival and reproductive capabilities of individuals.
2) *Demographic uncertainty* resulting from random events in the survival and reproduction of individuals in populations.
3) *Environmental uncertainty* due to unpredictable changes in climate, weather, food supply, and the populations of competitors, predators, etc.
4) *Catastrophic uncertainty* from such phenomena as hurricanes, fires, droughts, etc., which occur at random intervals.[37]

It is important to remember that, like the levels of biodiversity, these four classes interact. A disease (environmental uncertainty) could reduce the grizzly bear population in the North Cascades to six males and one female, the resulting unbalanced sex ratio (demographic uncertainty) could contribute to inbreeding (genetic uncertainty), and this could lead to few viable offspring and, eventually, extinction.

Although these four stochastic processes are difficult to model, one thing is certain—the magnitude of their effect on population dynamics increases as populations become smaller. A viable population is one that maintains vigor (health) and the potential for future evolutionary adaptation.[38] Populations face increasing risk as their numbers decline. At some point over time, all populations face some finite risk of extinction. If conservation biologists could discover, through the interactions of genetic, demographic, and environmental uncertainty, a minimum size below which a population's extinction was inevitable, conservationists and land managers would have guidelines to measure the success of their efforts.

Such numbers are not forthcoming. To understand why, we might follow conservation biologists and examine how each kind of uncertainty affects a population's chance of survival.

Biologists began their quest for viable population numbers by studying genetic uncertainty. Why genetics? There were several reasons. First, many of the scientists who first perceived the threat of reduced population size were geneticists and population biologists. Chief among these was Otto Frankel, a German geneticist who, as early as 1974, noted that the prime parameters for continuing evolution were the size of viable populations and the size of nature reserves.[39] Second, because genes are the basis of an individual's ability to track environmental change, there was a belief that some "genetic basement" number existed below which a population could not persist for very long. Finally, there was a relative wealth of data on the effects of inbreeding and loss of heterozygosity from small populations bred in captivity. When inbreeding occurs, reproduction and survival rates go down. Animals bear fewer total young, and the percentage of those that survive is less. Heterozygosity, where the genes on paired chromosomes are different at one or more locations, increases genetic variety. When heterozygosity is lost, an individual's ability to handle environmental stress is reduced and the mortality rate of the population increases.

Using empirical results from the work of animal breeders with captive populations and from laboratory experiments with fruit flies, biologists have determined that a population of about fifty individuals will not be harmed by inbreeding over the short term.[40] This number represents the genetic basement. But biologists note that populations of this size lose 25 percent of their genetic variation over a span of twenty to thirty generations. A manager could not count on maintaining many species at this reduced level for very long.

Five hundred has been used as the number of individuals needed for long-term population viability.[41] This is problematic. Over generations, genetic drift (random changes in the genetic pool of a given population) tends to reduce variation in a population, while random mutations may increase genetic diversity. Based on known average rates of mutation of 0.1 percent per generation, biologists have picked 50,000 generations as a long-term period for population viability. The result is that a deme of 500 should have enough staying power to cope with genetic uncertainty over time.

On the surface, the numbers 50-500 are the holy grail of conservation biology, a precise answer to the questions How low can you go? and What does it take to survive? But there are additional factors to keep in mind. The numbers 50-500 refer to effective population (Ne) size, not the census numbers familiar to managers and laypeople. A census is a gross count of all the members of a given group in a given area. It is the standard measure of wildlife biologists and game managers. Ne is a theoretical construct of population biologists, a complex concept meant to measure the reproductive contribution toward the next generation of individuals in a given group.[42] Consider the population of northern spotted owls dwelling in the Nooksack River watershed below Koma Kulshan. Researchers may produce a census of twenty-five owls, but perhaps only seven are of reproductive age and only five sexually active. The Ne of this deme would likely be five birds, not twenty-five. Another population of spotted owls in the Skagit River Valley might also number twenty-five individuals. Within this group, two owls may be contributing their genes to future generations; the Ne size would be two. The Ne will almost always be lower than the census size because not all individuals contribute equally to reproduction. A good guess is that census counts will be on average about four times higher than effective population.[43] For conservation biologists, this means that an Ne of fifty owls, at the bottom of the genetic basement, may actually translate into a census count of a few hundred birds. To maintain an Ne of 500 owls, several thousand individuals would probably require protection.

Most of the data that wildlife biologists have collected over the years on free-roaming populations has been census counts, and managers are not often adept at translating field results into theoretical Ne numbers. This problem is compounded by the fact that the vast majority of empirical research on Ne size has been conducted using farm animals and fruit flies.[44] Land managers and conservation biologists do not often review each other's assumptions; under the stress of endangered-species politics, this has caused confusion over the magic numbers of 50-500.

Thus the numbers 50-500 should be viewed with caution. And there is more to population viability than genetic uncertainty. Demographic and environmental stochasticity must also be taken into account.

The demographic character of any given group is almost never certain. Traits change in random fashion over time. No population of grizzlies is guaranteed a balanced ratio of males to females or young to old. Survival may be enhanced in a year of abundant food or depressed in a drought. Populations do not grow evenly over the years. Nor are bears distributed evenly across available habitat. They concentrate where tubers are plentiful, humans few, and cover is available, factors that change from place to place and year to year. Biologist Daniel Goodman has modeled demographic uncertainty and found that population-growth rates are the single most important variable in predicting whether a population will remain viable in the future.[45] The lower the growth rate, the more slowly average persistence time increases with population size. Goodman has also shown that unless a population is extremely small, fewer than 100 individuals, demographic uncertainty alone does not generate a great risk of extinction.

Because food supplies affect the growth and survival of grizzlies, environmental and demographic uncertainty are intimately linked. A year of little rain means less green growth, less nutrition, and harder times for many plants and animals. Droughts are random climatic events that make weather forecasting seem like child's play. Changes in predator and parasite populations may also be triggered by environmental change and alter demographic patterns. The barred owl has invaded the old-growth habitat of the northern spotted owl, which it seems to be able to outcompete. The barred owl has gained a foothold because clearcut logging has reduced forest patch size. The invading species thrives in smaller forest areas, while the spotted owl does not. The great horned owl, which preys upon spotted owls, also seems to be favored by fragmented forests. Spotted-owl demographics are changing as the bird reacts to a new competitor, increased predation, and the loss of the old-growth habitat it depends upon.

Managers might have been able to predict declining spotted-owl numbers in the face of a new predator and human alterations to habitat, but they would be hard-pressed to forecast a powerful windstorm or huge wildfire. Catastrophes represent the far end of the scale of environmental uncertainty. They may be common to age-old mountains, but they are singular events to humans. Both common environmental uncertainty and its unusual catastrophic forms are the wild cards in predicting viable population size. When the combined effects of demographic and environmental stochasticity are modeled, the message is clear: In "a variable environment any loss in population size proportionally increases the chances of population extinction. . . . To be very sure of retaining a species for very long may, depending on population growth rate and the degree of environmental variability, require either very large population sizes or numerous (sub) populations."[46]

The North Cascades grizzly may be living on borrowed time if a bird that went extinct in 1932 is any indication. Heath hen numbers declined drastically as New England was peopled during the nineteenth century.[47] By 1908, there were only fifty individuals left of what was once a common bird. Protected in a reserve, the population increased to 2,000 until a gale-driven fire killed many of the birds. This catastrophe was followed by several more. One winter was particularly frigid, an unusual northern goshawk migration increased predation, and finally, in

1920, a poultry disease reduced the population to below 100. A severe sex-ratio imbalance led to increased inbreeding, and in 1932 the last heath hen died. Environmental uncertainty doesn't disappear. In the southeastern United States in 1989, Hurricane Hugo ravaged the longleaf-pine nesting habitat of the endangered red cockaded woodpecker, and overnight roughly 12 percent of the population lost its nest-cavity trees.[48]

Conservation biologists have yet to produce a synergistic (combined) viable-population model that takes genetic, demographic, and environmental uncertainties into account. To simplify matters, biologists have constructed models that assume population characteristics and behavior (random mating, random distribution of offspring, equal numbers of breeding males and females, etc.) that are not often found in nature.[49] Current models are based on isolated populations, blanket effects across a homogeneous environment, and no biological adjustments to environmental change. In the real world, many populations consist of multiple subpopulations, no single disturbance affects all parts of a habitat equally, landscapes are varied and patchy, and animals almost always respond to environmental stress by switching food or moving to new locations. Nevertheless, the models constructed so far have racheted viable-population estimates up from the original genetics-derived 50-500 toward the low thousands or higher.[50] Most scientists today would concur with Russell Lande and George Barrowclough, who state that 500 is only "about the right order of magnitude" (that is, as low as 50 and as high as 5,000 individuals).[51]

Ignorance of the basic population dynamics of wild species compounds these problems. No one knows the census size of grizzlies or northern spotted owls in the Greater North Cascades, let alone their effective numbers, growth rates, or the ratio of males to females. It is extremely difficult and time consuming to collect such data under field conditions. (It is also expensive—the Fish and Wildlife Service spent almost $10 million on northern spotted owl research in 1990 alone.) Furthermore, a given species response to most of the viable-population variables is not constant from location to location. Each subpopulation of grizzly bears, from Greater Yellowstone to the Greater North Cascades, for example, will have a different viable-population size and needs to be studied separately. Once basic research on each subpopulation has been completed, the interactions (if any) between demes must be accounted for. In other words, "there is no single value or 'magic number' that has universal validity."[52]

Results from the most comprehensive viable-population computer simulation to date also dictate a conservative approach to any hard numbers. Working with twelve years of data on the grizzly bear in Greater Yellowstone, Mark Shaffer found that for an effective population of 50 animals (the current census size is about 200), expected time to extinction would be 114 years.[53] This led Shaffer and his coworker Fred Samson to conclude, "It seems clear from our limited work to date that realistic simulations will reveal that extinction probabilities are much higher and average times to extinction much shorter than is currently appreciated even for relatively large populations."[54] Updating grizzly-bear population predictions in 1991, biologists David Mattson and Matthew Reid found this to be true. Because of the predicted loss of key grizzly foods in the wake of climatic change caused by global warming, these biologists felt that "extinction over the next 100 years may be even more likely than suggested. . . ."[55]

Faced with fading magic numbers and the importance of environmental uncertainty, conservation biologists are gaining appreciation of the role of landscapes in population viability. Most species, as we have seen, are spread across ecosystems and not confined to a single subpopulation. The grizzly is a population of subpopulations with "movement between the groups much less than movement within them."[56] Scientists call this a metapopulation. The grizzly presents a landscape-scale problem in population biology. In the continental United States, each of the six subpopulations is isolated from the others and in the absence of migration must sink or swim in its own genetic and environmental pool. There may be 1,000 bears in the entire metapopulation, but as far as viable population theory goes, each distinct deme is dangerously low in numbers. Several of the U.S. subpopulations are completely cut off from the continental source pool of Canadian-Alaskan bear populations. It is important to remember that this situation is new for the grizzly. The animal did not evolve under the isolated conditions brought on so recently by human usurpation of its living space.

The theoretical bottom line is that, to remain viable, populations must number in the low thousands and/or be divided among numerous subpopulations able to exchange genetic material every now and then. The movement of individuals between demes within a metapopulation becomes critical. This flow depends on the distance between occupied habitat patches and the suitability of the intervening habitat. Currently, there is no way for a grizzly bear to travel from Yellowstone in northwestern Wyoming to Glacier National Park in northern Montana, let alone from the Selkirks to the North Cascades. It is not enough to live-trap a bear in Yellowstone and ship her to the Greater North Cascades; migrants must actually breed to be effective.

Metapopulation dynamics have led conservation biologists away from a search for universal magic numbers toward population-viability analysis, where all four classes of uncertainty are seen as interactive over space and time.[57] This view has forced land managers to become aware that, for many species, survival depends not only on habitat availability and quality but also on the spatial arrangement and flow of individuals between patches of living space. Fragmentation of natural ecosystems due to increasing human development has given us a new understanding of extinction as a process instead of an isolated event.[58] What may end in the loss of a plant, frog, or bird may have begun generations ago with some initial environmental stress that remained unchecked. The current lightning rate of human-generated habitat change is allowing us a view of extinction similar to that of Koma Kulshan's perspective on the birth and death of a single Douglas-fir. It has taken less than 150 years of logging and land clearing in the Pacific Northwest to "let light into the swamp" and reduce old-growth ecosystems and species to mere shadows of their former selves. No one knows at what point the northern spotted owl, grizzly, marbled murrelet, pine marten, and a host of other species began to feel the pinch. But each of these populations has been reduced in size and in proximity and flow between groups to the point of uncertainty. The story of the heath hen may be playing out once again at the opposite end of North America as ancient forest species lose their ability to adapt to a changing world.

Conservation science, through population-viability analysis, may soon be able to model how various chance events affect population persistence, but there is more to extinction than biology can tell. The loss of grizzlies and entire ecosystems takes place in time and always involves a certain degree of risk—one human activity, say logging the last stand of old growth inhabited by a pair of spotted owls in the Nooksack River Valley, may have more impact than cutting trees in a second-growth forest in the same watershed. Only society as a whole can determine how long the northern spotted owl should persist under what degree of risk.[59] Science cannot tell us much about the likelihood of the bird's survival without first being given the answers to two questions: What degree of risk is socially acceptable? and How does society define long-term persistence? These questions will be answered differently by different people depending on their environmental values. Mark Shaffer, in his Yellowstone grizzly simulations, chose as his answers a 99 percent probability of persistence for the bear over 1,000 years. Michael Soulé, in his textbook definition of a viable population, is willing to take greater chances with the future. Soulé selects a 95 percent probability of persistence over several centuries.[60] This still amounts to a low level of risk in relation to the values that may be expressed by nonbiologists. A Seattle storeowner might be satisfied with spotted-owl conservation efforts that would only provide a 50 percent likelihood of survival over the next twenty years. Someone who does not understand the implications of extinction might support no protection for the owl at all.

In the absence of societal consensus on these issues, scientists studying viable populations must select time frames and levels of risk on their own. Conservation biology is not only an applied science, it is also extremely value laden. Given the elusiveness of population-viability theory, most biologists believe that we should err on the side of greater security over longer periods. The need for caution is underscored when one realizes that, according to some scientists, the data on wild populations needed to flesh out viability models may take twenty or more years to collect.[61] The cost of gathering, analyzing, and incorporating such baseline data for many species, endangered or otherwise, has never been estimated.

Answers to the question How low can you go? may yet be as uncertain as a mountain snowstorm in summer, but two solid rules of thumb emerge out of population-viability analysis. There is no debate about a species gaining security as the number of individuals and the number of subpopulations increase, nor about viable populations needing habitat that meets their living requirements. This leads us to the second question that conservation biologists have pursued diligently: How big is big enough? This query has profound implications for a society that has yet to grasp that national parks and reserves do not automatically protect the plants and animals that depend on them.

The Decline of Nature Reserves

I never expected to have my life changed by a bar graph in a book and a table full of data hidden away in an obscure scientific journal—never thought formal figures would sway my soul to action. The early years of my education in ecology hardly prepared me for the biodiversity crisis.

As an undergraduate, when I was reading textbook descriptions of North American plant communities, I came across E. L. Braun's *Deciduous Forests of Eastern North America.*[62] Braun had surveyed the last large old-growth forests of the eastern and central United States in the 1930s and 1940s, just as the last remnants were being fragmented by the cities and suburbs, interstate highways, and cornfields that we have inherited today. I purchased this out-of-print hardcover on my shoestring student budget for the black-and-white photographs of old-growth American chestnut, beech, and sugar maple that I knew I would never see. I was less enamored of the data Braun presented detailing her investigations. I skipped over the tables and graphs for the pictures and descriptions of ancient forests. I objected to ecosystems being reduced to so many numbers through quadrants, transects, and the host of quantitative sampling techniques that community ecologists employ to describe ecosystems. I retained this aberrant predilection through college and a master's program, and was not prepared for the impact of William Newmark's provocative article on the legal and biotic boundaries of national parks, which I read in 1986.[63]

What Newmark published was elegant if somewhat simplistic. First, he calculated the legal area of eight of the largest parks or groups of reserves in western Canada and the United States. He then selected several native mammals from each park that were known for their large home ranges (grizzly bear, mountain lion [*Felis concolor*], wolverine

[*Gulo luscus*]) and extracted from published field accounts the amount of territory necessary to sustain one individual of each species. Newmark then multiplied the average home-range sizes by the numbers 50 and 500 to obtain a rough indication of how much habitat an entire viable population might require over the short and long terms, respectively. Finally, he compared the legal sizes of each park with what he described as the biotic boundaries of each area.

Newmark's results were extremely disconcerting. Yosemite National Park was four times too small for fifty wolverines. Olympic National Park, even at almost one million acres, was not sufficient space for 500 mountain lions. Even the largest area in Newmark's study, a combination of the biggest parks in the Canadian Rockies, was six times too small to harbor 500 grizzly bears over the long term. Along with many of the biologists who read the article, I was shocked into an awareness that, even if these rough calculations were considerably off the mark, national parks were not going to protect wide-ranging carnivores much into the future.

Not long after encountering Newmark, I read a second article that covered the same ground.[64] What struck me was not the text but rather a single bar graph (figure 2). The figure charted the area of ten of the most extensive and undeveloped public wildlands in the continental United States. Against these "conservation networks," the authors, ecologists employed by the U.S. Forest Service and National Park Service, graphed a curve based on actual census data on large carnivores known to live in the networks. The curve combined data for seven species in a single biostatistical, computer-generated figure. The figure suggests that none of the ten areas is large enough to support a census population of 500 animals. Only the huge Selway-Bitterroot wildlands of central Idaho (the largest roadless area in the lower forty-eight states) provides enough space for what would likely translate into an effective population for native species. The North Cascades conservation network doesn't even come close to doing this. What the bar graph shows is a future bereft of large North American mammals.

To construct their figure, Hal Salwasser and his coauthors used population data reported in a 1983 study by Dr. Christine Schonewald-Cox at the University of California, Davis, Park Service Cooperative Re-

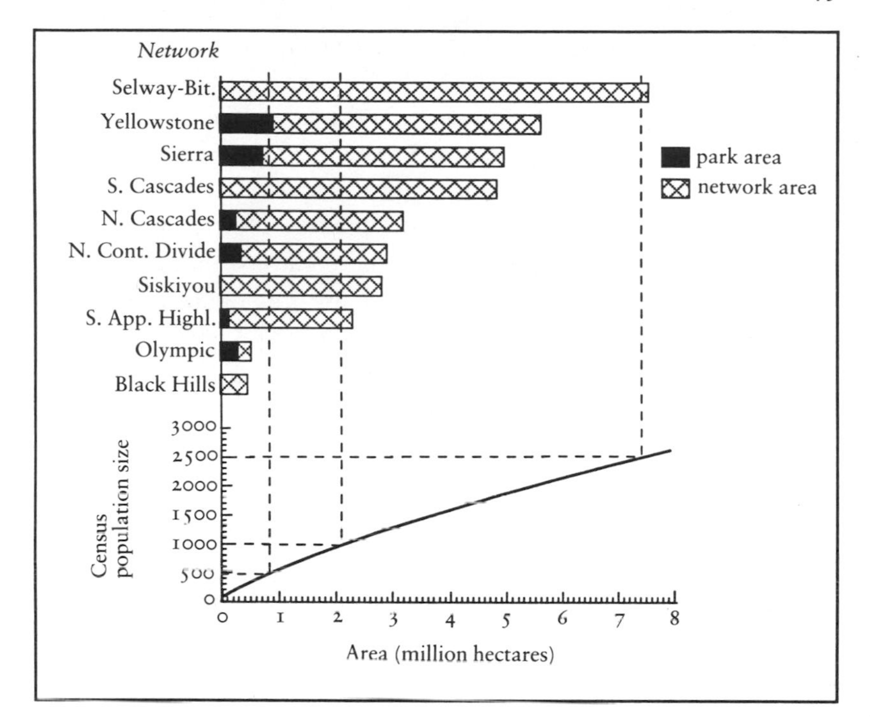

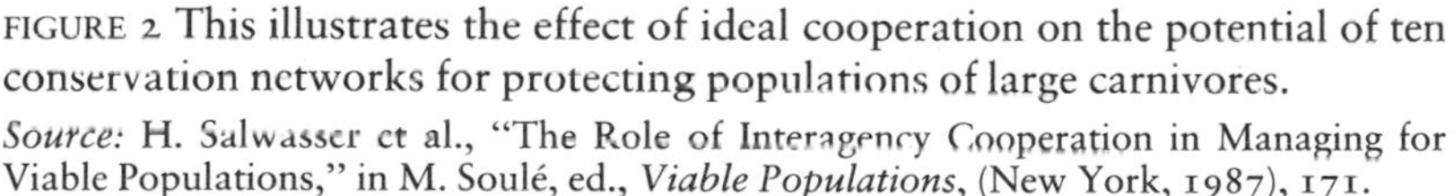
FIGURE 2 This illustrates the effect of ideal cooperation on the potential of ten conservation networks for protecting populations of large carnivores.

Source: H. Salwasser et al., "The Role of Interagency Cooperation in Managing for Viable Populations," in M. Soulé, ed., *Viable Populations*, (New York, 1987), 171.

search Unit.[65] Searching for the statistical reasoning behind Salwasser's work, I came upon an unsettling quote used by Schonewald-Cox to summarize her own results: 55 percent of U.S. national parks would "for most vertebrates and some long-lived plant species . . . provide little protection beyond the life of individuals of a few dwindling generations that presently constitute the populations."[66] Even in the larger parks (250,000 acres or larger) such as Olympic National Park, Rocky Mountain Park, and the Grand Canyon, "the prospect for the large, slowly reproducing and highly specialized species (large carnivores and many of the ungulates) is dim." With only 10 percent of U.S. national parks protecting this much territory and many of these clustered in Alaska, it is not difficult to understand the implications of inadequate reserves for protecting viable populations of large mammals.

How much trust can one place in computer-generated population curves, bar graphs, and simplified relationships between the area of a given national park and the average home ranges of several carnivore species? Newmark, Salwasser, and Schonewald-Cox would all admit that their research results are speculative. Unfortunately, the first empirically derived extinction-probability study, reported by Gary Belovsky of the University of Michigan, confirms their theoretical conclusions.[67] In estimating the minimum-area requirements for mammals of varying body size, Belovsky noted that, in general, larger mammals exhibited lower growth rates and population densities than mammals with small bodies. Using data from species living on mountain ranges in Nevada, he projected the survival probabilities of large carnivores to be a dismal 0 to 22 percent over a hundred years. Large herbivores such as elk and deer would likely survive longer. Small mammals did not seem to be at risk.

The initial studies that attempted to answer the question How big is big enough? all depict a dangerously uncertain future for many large mammals. Within a century or less, black bears, grizzlies, wolves, wolverines, mountain lions, and possibly elk species, bighorn sheep, bison, and moose populations may all be gone or shrinking toward oblivion. Conservation biologist Michael Soulé sums up these studies bluntly: "Most [viable populations] will be so large that it will be impossible to contain this many individuals of a large animal in reserves and sanctuaries of modest size (up to thousands of km^2)."[68] By the time my children are as old as I am now, there may be no place left for a grizzly bear to hide.

Effects of Habitat Fragmentation

It is spring and I am searching for a landscape with ecological integrity. I am looking for a low-elevation valley where I can hike in old-growth Douglas-fir forest for a few days with my students. Such a valley is not easy to find in Washington State. In his 1940s inventory of Forest Service roadless areas, Aldo Leopold suggested that if an area could not sustain a two-week pack trip, then it was not suitable for wilderness.

My criteria are less stringent—I would be satisfied to find a North Cascades watershed where we could hike four to five days round trip.

According to the definition published in 1986 by a Forest Service research team headed by Jerry Franklin, Douglas-fir forests take about 200 years to acquire characteristic old-growth conditions.[69] These include a minimum of six to eight large-diameter trees per acre, a multilayered, uneven canopy with plenty of broken tops and snags, and ten to fifteen tons of downed logs and deadwood per acre. To find an old-growth hike, I have to locate a forest ecosystem that has developed uninterrupted for a minimum of two centuries.

Another aspect that colors my search is the number of days I wish the class to camp in ancient forest. I need a forest extensive enough to explore for five days without walking into second growth and clearcuts, and without gaining enough elevation to climb into montane forests above 4,000 feet.

In the 140 linear miles from the Canadian border to Mount Rainier National Park, I have discovered only eight westside valleys with hiking trails that come close to meeting these criteria. Only six watersheds, Big and Little Beaver creeks, Thunder Creek, and the Chilliwack, Suiattle, and Whitechuck rivers, contain ten or more miles of old-growth hiking. Entire subregions of the North Cascades have no sustained old-growth hiking at all. On the north side of Mount Baker, an area of national forest encompassing over 300 square miles, almost every backpacking trailhead begins above 3,000 feet and proceeds quickly to the subalpine high country.

The paucity of low-elevation old-growth presents a problem to a teacher trying to find an optimal educational route. But consider the difficulties it poses for the northern spotted owl, which requires regional old-growth ecosystems for survival. Research has shown that each pair of spotted owls in the North Cascades needs thousands of acres of forest from which to get their living.[70] This is ecosystem-level space over a portion of a particular watershed. Multiply the ecosystem needs of two spotted owls times the estimated size of a viable population and the species runs into risks not unlike those of the wide-ranging mammals that also appear to be threatened by their need for large landscapes. The protection of the spotted owl requires millions of acres of old-growth ecosystems. Put in simple terms, as the Pacific Northwest's ancient forest goes, so goes the northern spotted owl.

If the difficulty of finding an ancient-forest backpack route is any indication, the owl is in trouble. The reason is logging. Turn-of-the-century land surveys suggest that about 85 percent of western Washington and 90 percent of western Oregon were once old-growth forest.[71] Not all of this acreage would have met today's ecological definition of old-growth. Ecologist Elliott Norse estimates that about 19 million acres is a "plausible starting point" for defining the regional extent of ancient forest prior to industrial logging and land clearing.[72]

Cutting began on the most productive low-elevation sites on private lands and proceeded up the river valleys and into the mountains. By 1946, almost 70 percent of the original forest had been cleared. National-forest logging began in earnest after World War II to feed the postwar housing boom. By the beginning of the 1990s, most privately owned ancient forests had been fed into mills. Today, with the twelve westside national forests still containing 30 percent of the commercial timber found on *all* Forest Service lands nationwide, it is easy to see why the timber industry's primary goal has become the logging of the last publicly owned old-growth.[73]

How much old-growth is left? Even after two major studies, no one knows precisely. There are several reasons for this. Chief among them is the fact that any answer is subject to political manipulation. A brief history of attempts to settle the question proves instructive.

At first, the Forest Service ignored the old-growth definition provided in 1986 by Franklin's agency-commissioned research group. To use the strict ecological definition would have been politically disastrous for the simple reason that if Congress and the public found out so few acres remained, the pressure to reduce logging would have been severe. Instead, in the forest plans published in the late 1980s under the National Forest Management Act, each westside forest used its own definition for ancient forest, and most of these lacked scientific rigor. The largest national forest in the North Cascades, the Mount Baker–Snoqualmie, used "mature sawtimber with diameter breast height 21 inches or greater" to calculate acres of old-growth.[74] No Forest Service plan employed the ecological definition published by Franklin's team.

The differences between old-growth estimates based on the ecological definition and nonscientific Forest Service accounts are large. To challenge the Forest Service, the Wilderness Society hired ecologist Peter Morrison to do what the agency had not done: Use the Franklin definition. By 1988, Morrison could show that Forest Service estimates for six of the westside forests were inflated by 55 percent.[75] While the six forest plans stated that 2.5 million acres existed, Morrison could find only 1.1 million acres. Facing Morrison's published results, the Forest Service had to commit to doing its own ecologically based survey. Both the Wilderness Society and the agency presented comprehensive findings in 1991, and still politics fueled discrepancies in the studies.[76] The Forest Service found 4.3 million acres of ancient forest but did not take into account all parts of the scientific definition. Morrison still found less than half of what the Forest Service claimed was there. And to complicate matters, the Forest Service, by stretching the definition of "protected," concluded that 68 percent of remaining ancient forest land was safe from logging. Assuming protected to mean *permanently* closed to logging, the Wilderness Society study found that only 54 percent of ancient forest land met this standard.

Since neither study stopped any logging and trees have continued to be cut, both sets of figures were out of date before they were publicized. As of early 1992, it is likely that less than 10 percent of ancient forest ecosystems remain. This figure is one overall measure of the predicament of the northern spotted owl and the pressure on an increasingly endangered ecosystem type. It also indicates the level of habitat fragmentation that humans have wrought on Pacific Northwest landscapes.

According to conservation biologists Bruce Wilcox and Dennis Murphy, "habitat fragmentation is the most serious threat to biological diversity and is the primary cause of the present extinction crisis."[77] Many other scientists would agree. Remember that spotted owls, like grizzlies and wolves, inhabit ecological landscapes, not abstract acres of old growth, national parks, or reserves. Reserve-size studies, conservation-network analyses, and ancient-forest inventories all share a shortcoming—they tend to emphasize gross area over functional habitat. Referring to gross acreages of park- and forestland as if they were pristine blocks of wilderness, these studies fail to take into account the actual ecological conditions on the public lands they inventory. A century of logging, roadbuilding, and other development has cut ecosystems into bits and pieces of varying sizes and shapes throughout the North Cascades and the United States. Some of these fragments remain healthy, some are compromised, and some have become ecologically dysfunctional.

What are the landscape-scale dimensions of habitat fragmentation and how do they affect extinction? There are two components of habitat fragmentation; both can contribute to the loss of species.[78] *Habitat loss* occurs when the total area of an ecosystem is reduced. A 100-acre clearcut removes a hundred acres of old-growth forest. *Habitat insularization* increases as habitat loss proceeds. The remaining fragments of ecosystems become increasingly isolated from one another as large expanses of living space are transformed into a number of smaller patches. Habitat loss destroys, reduces, and subdivides populations, and all of these processes increase the probability of extinction. Insularization impedes species movements from fragment to fragment, which decreases immigration and dispersal rates. Again, the risk of extinction rises.

The effects of fragmentation are complex. They occur over time, and as long as patches of habitat remain sufficiently large, few species are lost. At some point, however, threshold levels are reached and those species that require large patches begin to vanish. Today, in the North Cascades, it appears that insularization has reached a level where the northern spotted owl and other species dependent on old-growth are balancing on the brink.

Conservation biologists have identified several specific ways in which animal populations can be affected by subdivided landscapes. As reserve-size studies indicate, a habitat patch must be at least as large as the home range of the species that live within it. Spotted owls in the North Cascades cannot exist in old-growth fragments less than about 4,000 acres unless they can fly to a neighboring patch of ancient forest. If they cannot negotiate the intervening habitat because of the distance between patches, say, or the presence of predators, it makes no difference whether an adjacent stand exists. The birds will be functionally isolated.

If a habitat patch accommodates a species' spatial needs, it must also contain all the resources a bird or bear requires to survive. Many animals depend on resources found across several different kinds of ecosystems, what biologists refer to as habitat heterogeneity. Grizzlies, mountain lions, and many wide-ranging creatures are automatically at risk in fragmented landscapes owing to their taste for variety. If a patch becomes a prison, these species must be well provisioned or they will begin to disappear.

Those animals that depend on undisturbed forest interiors are also at risk. Just as salaried employees learn the difference between gross income and net take-home pay, conservation biologists have come to distinguish ecologically functional habitat from the total area of a forest fragment. The amount of "ecological income," or usable living space in a habitat patch, is usually less than the gross area of the patch. Why?

Native species are not subject to taxes, but some are susceptible to edge effects. Edges, or ecotones, exist where two different plant communities or successional stages meet. This usually results in a greater number of species being present because several different ecosystems are juxtaposed. Consider the contrast between a sunny clearcut and an ancient forest interior, then the relationship between the two. The environmental conditions of the open area, high soil and air temperature, low humidity, intense light, and greater wind speed penetrate some distance into the forest, changing the microclimate inside the perimeter of the patch. Seeds from the clearcut rain into the patch, perhaps altering plant-species composition. As you move away from the edge of the patch toward the center, these changes become less apparent until, deep within the forest, undisturbed conditions prevail.

Wildlife biologists have traditionally viewed edges as positive influences on species and ecosystems. As the saying goes, The more edge the better. But recent studies of the effects of habitat fragmentation prove that not all species benefit from edges. Forest songbirds that require large breeding territories in forest interiors have shown long-term population declines in highly fragmented eastern deciduous forests.[79] Nest predation and parasitism increase dramatically near the inside edges of habitat patches.[80] Much of this is because of increased populations of raccoons, skunks, opossums, and other middle-sized omnivores that

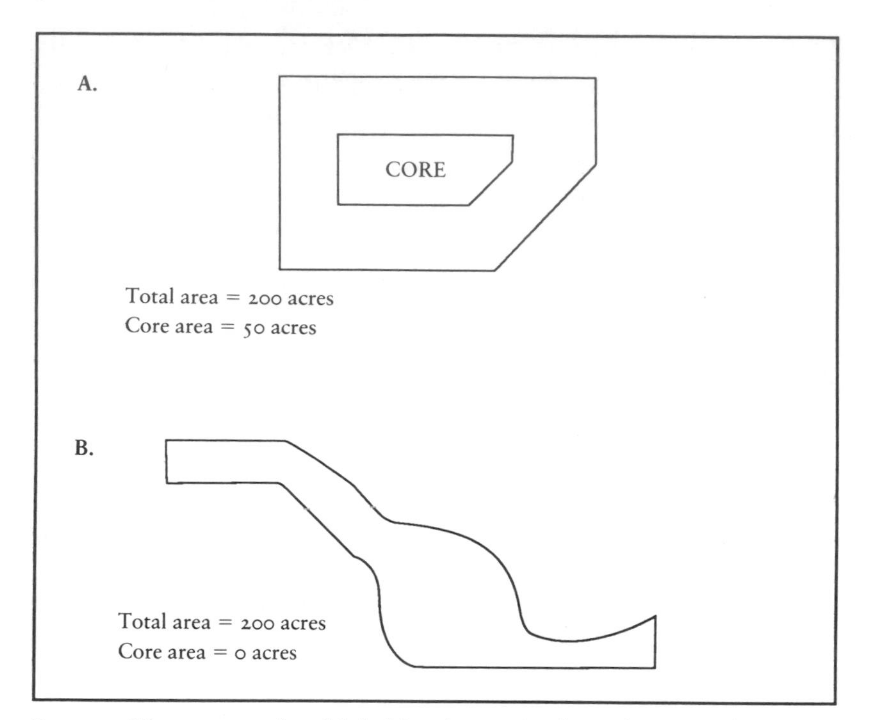

FIGURE 3 The amount of usable habitat for species dependent upon forest interior conditions varies with shape as well as size of the habitat fragment. Note that both A. and B. are the same size, but edge effects prevail throughout B. due to its narrow shape.

thrive in fragmented landscapes and invade native-habitat patches. Though data is sparse from old-growth ecosystems, evidence from other regions suggests the loss of forest-interior conditions for surprising distances into fragments. Research by David Wilcove and others has shown that songbird predators may penetrate habitat fragments as far as 30 to 1,600 feet.[81] The developing rule of thumb in the Northwest is that edge effects can generally be expected up to 400 feet within a forest patch, roughly the height of two canopy trees. This means that a patch of ancient forest below twenty-five acres in size is all edge and provides no interior habitat for old-growth-dependent species. A 200-acre circular stand will only be 25 percent ecologically functional for interior species.

What about a similar-sized habitat patch that buffers the banks of a creek as it tumbles down a mountainside? Because of its linear shape, a narrow forest fragment may have no interior habitat at all (figure 3). Shape is as important as size in the ecology of patchy landscapes.

By definition, deep-forest conditions are absent from the matrix of clearcuts, plantations, and second-growth stands that surround old-growth fragments. Managed stands bear little resemblance to natural forests.[82] Take a walk in a thirty-year-old stand of Douglas-fir. The developing canopy, sixty feet high, filters light through a uniform green crown. The clearcut that brought this new forest into being removed most of the snags. The slash burn that followed incinerated the large down logs that once littered the forest floor. Forest ecologists concerned over such structural changes have modeled the loss of coarse woody debris over time.[83] At the end of the first cycle of cutting, when ancient forest becomes a managed stand, 70 percent of this material is lost. Only 6 percent is projected to remain after the second cut. Without the large trees, standing snags, and dead and downed logs that make ancient forests unique, much of the structural diversity that native species depend upon is lost. Compared with natural forests, managed stands support fewer kinds of breeding birds, small mammals, and amphibians, and greatly reduced numbers of birds and amphibians. Managers cannot "remodel" old-growth forests without compromising native diversity.

The managed stands between old-growth fragments do not soon become ancient forest. In northwestern California, a century of clear-cutting has reversed the historical ratio of mature forest to second growth. Before logging, about 74 percent of the region was mature or old-growth forest.[84] Now, 63 percent is second growth. The trend is the same throughout the rest of the Northwest, and no private timber company or federal agency has plans to "grow" ancient forest.

Peter Morrison is searching for the future of the fragmented forest. He wants to know how size, shape, and edge effects combine with managed landscapes to compromise the ecological character of old-growth patches. When it comes to old-growth inventories, unlike the Forest Service, Morrison is a "splitter," not a "lumper." He is capable of making the appropriate ecological distinctions between different kinds of old-growth habitat. His survey of the Mount Baker–Snoqualmie National Forest is revealing on this point. In the forest plan the Forest Service, using outdated definitions and data, states that there are 667,000 acres of old-growth woods left.[85] Morrison the scientist, armed with Franklin's ecological definition, can find only 297,000 acres.[86] That is only the beginning. Morrison the *conservation biologist* notes that 24 percent of this total is in stands of less than 400 acres surrounded by managed landscapes—clearcuts, plantations, roads, and second growth. These forests, to some degree, are compromised by the effects of habitat fragmentation. They are less capable of sustaining owls, bears, and forest songbirds. "From the point of view of managing shredded habitat," says Morrison, "you need the kind of differentiation our project uses. But the Forest Service and the timber industry keep trying to pool it all in one big mass as if it's all the same."[87]

The forests of the Pacific Northwest have not been created equally. Some patches are dysfunctional, and some are still part of their original ecological matrix—whole ecosystems spread across unfragmented landscapes. There is virtually no old-growth left on private lands. For better or worse, most of the remaining ancient forest is public forest, the key to the future of native biodiversity in the North Cascades. This pattern holds throughout the United States. Almost by default, federal lands are the last bastions of wild nature in America.

Habitat fragmentation does not operate in the past tense. On all the national forests in western Oregon from 1972 to 1987, the area in managed forest increased twofold and forest-interior habitat decreased by 18 percent.[88] More timber was cut in the national forests of the Pacific Northwest in 1990 than in any year in history. Using Forest Service data to monitor several "indicators of biological impoverishment," conservation biologist Reed Noss has found that over the next fifty years in westside national forests, old growth will decrease by 31 percent, road density will increase, and less than 10 percent of forestland outside of designated wilderness will remain roadless.[89] If present trends continue, in one generation only 6 percent of the world's most biologically diverse temperate-zone forest will remain. That may be too little too late for the forests of the future—nothing that scientists understand about biodiversity suggests that 94 percent habitat reduction bodes well for ancient-forest species and ecosystems.

Natural Disturbances and Patchy Landscapes

> 4 August 1990. Marcy was the first to smell smoke this morning as we splashed through the ford at the Baekos Creek crossing. I thought it might be from the trail crew cookfire by the Whitechuck River back down the trail but that was too far away and there would be no reason for them to have a fire going in the middle of the day in this dry weather. And the wind was blowing from the wrong direction . . . Now we're up above 6,000 feet and we can see where it is coming from—three separate smoke plumes rise from the White River drainage east of the Cascade Crest, maybe six miles away. One fire is quite a bit larger than the other two. Smoke is filling the valley, rising, spilling over the divide in all directions. Glacier Peak is screened as if a gauze curtain was hung from the uppermost icefalls. Must have been the lightning storm we had two nights ago. Green mountains in the dry season burn. Every now and then, through binoculars, I can just make out a shaft of flame climbing a tree towards heaven. No rain. Dry woods. Midsummer. Fire weather.

As the rest of my journal entries over the next five days attest, we lived with fire on that backpack. The forests of the Pacific Northwest have experienced this most widespread of natural disturbances for thousands of years. Lightning strikes, a fire explodes, and the march of forests toward the future is given a new direction.

The traditional view of old-growth dynamics in westside forests begins with a charred stand or jackstrawed pile of treetrunks tossed down by wind. Forests in the Northwest, after such catastrophic disturbances, are replaced by young communities dominated by fireweed and a host of fast-growing herbs and shrubs. Even-aged Douglas-firs soon become established and rapid tree growth persists from canopy closure (20 to 50 years) until stand maturity (80 to 100 years). Second growth becomes mature forest. In the absence of another disturbance, maturing forest continues to develop old-growth structural characteristics—large, dominant canopy trees, snags, and down logs—until, after some 200 years, ancient forest conditions prevail. Succession may continue for centuries as shade-tolerant trees such as western hemlock slowly replace the original Douglas-firs in a predictable process of change.

But the picture is not so simple. Wildfires are not orderly, and neither are windstorms, floods, avalanches, and other large-scale natural disturbances. The effects of disturbances on landscapes hold answers to the question How big is big enough?

Across the North Cascades, the Pacific Northwest, and western North America in general, fire structures ecosystems. But because fires differ in size and intensity, only a portion of a landscape is affected at any one time. Succession in old-growth ecosystems is not predictable or homogeneous much beyond the short term. Instead, natural forests are patchwork quilts that record the history of disturbances both large and small. Catastrophic or stand-replacement fires do occur along the lines of the traditional model. The mean period, however, between such conflagrations varies from 465 years in Mount Rainier National Park to anywhere from 150 to 276 years in the western Oregon Cascades.[90] In their study of 800 years of fire history in old-growth stands in Oregon, Peter Morrison and Fred Samson observed that some sites burned every twenty years, while others were fire free for 500-year intervals. Large fires were infrequent—86 percent of the burns they recorded covered less than forty acres. Intensity varied widely as well: Understory burns occurred at least as often as high-intensity fires. These results fit into a general pattern across the westside from the North Cascades to southwest Oregon, where natural fires are more frequent and less severe. Morrison and Samson also discovered that, counter to the conclusion of the stand-replacement model, the structural identity of many old-growth stands (large trees, down logs) and snags survives repeated burns.

Fire may be the most important natural disturbance in the Northwest, but it is not the only one. A black bear rips open a log to feed on a termite colony. An ancient red cedar topples over, taking down neighbors fifty yards away and opening a gap in the canopy through which sunshine pours. There are patches within patches as ecosystem processes (tree fall, wildfire, wind throw) create ecosystem patterns (a down log here, a forty-year-old forest there, old-growth across the watershed). Patch conditions are superimposed on ever-changing environmental conditions such as soil type, rock outcrops, available moisture, topography, and the direction a slope faces, north or south. When ecologists combine the effects of disturbances and baseline environmental conditions to build a dynamic picture of landscape diversity at all scales, they see a " 'shifting-mosaic steady state' of patches in various stages of recovery from disturbance."[91] Species diversity is ultimately connected with landscape patterns. According to this view of nature, old-growth means more than just big trees and chronology, "x-amount of trees at least two hundred years old," and becomes a complex weave of owls and ancient trees, snags and light gaps, fire and rain.

If people wish to protect species diversity, they must pay attention to the landscape patterns and processes that shape edges and habitat patches. Given what is already known about fire, this may be a daunting task. The argument goes like this: A park or reserve must be large enough so that only a small portion of it is disturbed at any one time. Otherwise, at some point the reserve will be subject to a "catastrophic" disruption that will alter the entire suite of ecosystems within its borders. There will be no source within the reserve itself for animals and plants to recolonize the disturbed area. This is the landscape-level version of the island dilemma, a regionwide implication of the biodiversity crisis with human development and habitat fragmentation on the upswing.

In 1978, ecologists Steward Pickett and John Thompson published a paper that offered a provocative solution to managing reserves on the same scale as nature. They suggested that parks include a "minimum dynamic area," or the "smallest area with a natural disturbance regime which maintains internal recolonization sources, and hence minimizes extinctions."[92] Reserves of this size would always be able to replenish themselves from areas inside their boundaries that have not been burned, blown down, flooded, or otherwise disrupted. They could "manage themselves" into the future.

The perceptive reader may note similarities in the jargon that population biologists and landscape-oriented ecologists employ in discussing populations and disturbances. There are minimum viable populations and minimum dynamic areas. Just as conservation biologists recommend margins of safety in favor of grizzly bears and rare orchids to offset the effects of genetic, demographic, and environmental uncertainty, so landscape ecologists suggest caution when considering disturbances in the design of parks and reserves. Shugart and West speculate that to maintain landscapes, "quasi-equilibrium" areas should be fifty to a hundred times larger than the largest known disturbance patch in the region.[93] This sounds simple enough: The larger the disturbance, the larger the reserve and then add a bit more. But that little bit is large indeed when one considers the huge area that some wildfires burn. The largest fire in the North Cascades this century burned over 40,000 acres of the upper Skagit watershed. The 1987 Silver Fire in southern Oregon played over some 200,000 acres. Recall the Yellowstone fires of that same year. For months they burned over an area considerably larger than the park itself. If these fires represent the largest disturbance that might pattern the Yellowstone landscape and you factor in Shugart and West's safety margin, the resulting reserve would dwarf the size of any area now protected in the world.

Maybe Shugart and West aim too high. A park the size of Yellowstone may, in fact, approach the area necessary to sustain some northern Rockies landscapes.[94] Little empirical research has been done on the question. Ecologists do not yet know enough to speak with much certainty. But two things are clear: Huge wildfires are not uncommon throughout western North American ecosystems, and to recall Christine Schonewald-Cox's research, few U.S. parks appear large enough to contain minimum dynamic areas.

Reserves smaller than Yellowstone or the North Cascades are not useless. Most animals do not range widely, plant species move slowly, and disturbances do not usually consume regional landscapes. But there is no escaping the fact that small reserves—if isolated and surrounded by unsuitable habitat—will have problems with population and ecosystem viability. If landscapes are patterned by natural disturbances, then conservation strategies should be expanded beyond the current focus on species and populations. There are two basic approaches: manage landscapes as wholes, using viable-population studies combined with research on minimum dynamic areas, and/or protect what native habitats remain by reconnecting them. Both of these have led some conservation biologists to look at national parks and forests as parts of greater ecosystems.

Greater Ecosystems and Biological Corridors

What happens when you radio-collar a grizzly bear? In 1959, Frank and John Craighead found out. These two biologists pioneered the use of radio telemetry in tracing the movements and locations of grizzlies in Yellowstone National Park.[95] Small radio transmitters were implanted in plastic collars and fitted to bears that had been momentarily captured and drugged. This allowed the Craigheads to monitor population movements, home ranges, and a host of other demographic data across the Yellowstone landscape. After twelve years of tracking grizzlies, the Craigheads were the acknowledged experts in the field. Then a conflict with the Park Service over bear-management policy put an end to their research.

In frustration, the Craigheads left Yellowstone and their grizzlies. But not before Frank Craighead had reached one startling conclusion: The park was not big enough for the bear. The population roamed far beyond the 2 million-plus acres of national park to meet their requirements for food, space, and solitude. Each adult female needed about 64,000 acres for her home range; if these territories were not overlapping, that worked out to enough room for maybe forty bears.[96] The Craigheads departed in 1971 when viable populations, habitat fragmentation, and disturbance regimes were hardly household words. Nevertheless, they could see the future in the quickening pace of Forest Service logging, oil and gas exploration, Park Service tourist development, and the rise of second-home vacation retreats. When he published his research results in 1979, Frank Craighead called for the recognition of a greater ecosystem in and around Yellowstone National Park. He suggested that the boundary of a Greater Yellowstone, based on the primary habitat needs of the bear, might be 5 million acres.

There was little response to Frank Craighead's proposal from the Park and Forest Services, the two federal agencies with authority over most of the land in the region. The political problem was obvious—neither agency felt bound to place the grizzly at the top of its list of priorities. Private landowners in Greater Yellowstone feared the loss of their property. To this day, the federal agencies are uncomfortable with the idea of greater ecosystems. Twelve years after Craighead's proposal, Yellowstone is only one of three areas in the United States where the term is much employed. The Greater North Cascades and the southern Appalachian highlands ecosystems are the others. Why aren't other interrelated ecosystems regarded as greater ecosystems?

There are scientific as well as political problems with the concept of greater ecosystems. First, the scientific definition of ecosystem, the interrelationship between species and their environment, is scale-neutral. Setting specific ecosystem boundaries depends on the question being asked, the particular species present, and the natural disturbances in operation. Craighead's greater-ecosystem boundary depends on the spatial requirements of the grizzly; the needs of Rocky Mountain elk or the pattern of a natural-fire regime would result in other borders. For example, estimates of the size of the Greater Yellowstone ecosystem range from 6 to 18 million acres. Estimates for the Greater North Cascades range from 6 to 8 million acres. It should also be noted that the greater-ecosystem concept does not automatically focus on large terrestrial mammals. As a group, large land animals make up only about 0.02 percent of all North American species. In the context of overall numbers, however, some of these species may be disproportionately valuable to conservation efforts. Conservation biologists are fond of using an umbrella analogy to describe those animals that roam over so much territory that to protect their habitat would likely conserve the habitat of species that need less space.

Because of fuzzy definitions and political resistance, most of the debate over greater ecosystems is taking place in scientific journals and behind closed doors at agency management meetings. Ecologist Patrick Bourgeron makes clear that greater ecosystems are not self-evident; what is to be "protected," and what is meant by a "representative" natural-disturbance pattern, ecosystem type, or home range, must be defined clearly.[97] With viable-population theory and disturbance ecology still developing, precise definitions based on specific field data are the exception rather than the rule. There is even some question as to whether regional landscapes are ever in any kind of long-term equilibrium that could be designed into management plans.[98] There is probably no such thing as a "natural" type of vegetation for any given place. Ecologist Douglas Sprugel provides a summary: "Because chance (disturbance) factors and small climatic variation can apparently cause very substantial changes in vegetation, the biota and associated ecosystem processes for any given landscape will vary substantially over any significant time period—and no one variant is more 'natural' than the others."[99]

The current boundaries of the Greater North Cascades ecosystem reflect our imperfect knowledge since there has been no peer-reviewed study of the outline of the region (see figure 1, page 10). Ecologically savvy grassroots activists are responsible for the work accomplished so far. The eastern boundary generally follows the Okanogan River valley that separates the edge of the mountains from the lava plateaus and sagebrush of the upper Columbia basin. The southern border results from a change in the geology and soils of the Cascade Range. Considering the ecology of Pacific salmonids, one can argue that salt water is the only defensible western margin for the Greater North Cascades. But the Puget Sound lowlands are certainly not in the mountains, and the presence of most of the human population of Washington state makes for certain difficulties. The western boundary given in figure 1 follows the foothills of the Cascades. The northern border is *terra incognita*: We know little about the biogeography of southern British Columbia and the population centers of large mammals (grizzly bears, wolves, lynx, etc.), and we know even less about the relative health of ancient-forest ecosystems, given the extensive logging that has occurred.

Though it is still being discussed, the greater ecosystem concept more closely matches what scientists are learning about biodiversity than any other current management framework. It incorporates the play of wildfire and other natural disturbances as they shape ecosystems, and it includes the requirements of individual species for living space. The main stumbling block is that the concept renders current notions of land-management practice and politics obsolete.

If conservation biologists are having difficulty providing a scientific basis for managing landscapes, are they having more success reconnecting regional fragmented habitat? If a park the size of Yellowstone or the North Cascades cannot function on its own, is it possible to somehow connect it with other reserves?

Biological corridors may be the future for protecting biodiversity, and most biologists advocate them as part of an overall conservation strategy. Corridors are connecting strips or swaths of suitable habitat that allow animals and plants to move between otherwise isolated territories. The idea is that if bears or owls do not have their needs met in a small, old-growth fragment or a medium-sized reserve, they could survive the uncertainty of small populations by reproducing with individuals from other regional reserves. Similarly, if a wildfire completely burned a 3,000-acre park, animals and plants from linked reserves could recolonize the blackened area without having to negotiate clearcuts, farmland, or other unsuitable habitat. If it proves impossible to maximize the protection of greater ecosystems, the next best strategy is to keep routes of movement open between existing parks and reserves.

The problem with corridors is that conservation biologists know more about the negative consequences of severed habitat than they do about the specific benefits of reconnecting habitat patches. There is debate as to the efficacy of knitting ecosystems back together. The main advantage would be enhancing genetic exchange and thereby decreasing threats to small populations. But linked reserves could also facilitate the spread of disease and "contagious catastrophes" such as wildfire.[100] As with many reserve-design theories, these arguments cannot be resolved without understanding the particular needs of specific organisms that might benefit from corridors as well as the ecological factors at play in specific places. There is some data on animal movements in forest patch/agricultural settings, but regional-scale research from relatively undisturbed ecosystems is lacking.[101] But lack of information and disagreement among biologists should not take the luster off the idea of biological corridors. One must remember Koma Kulshan's view: The original landscape was interconnected.

A New Ecological Image of Nature

May you live in interesting times, goes an ancient Chinese curse. As if wildfires, pest outbreaks, and hurricanes were not enough, humans have raised the stakes throughout the world with their own unique brands of change. The perversity is that while most individuals would like to arrest change, the sum total of human activity has led toward greater instability in natural ecosystems. Conservation biologists describe some of the effects of human activity as extinction and habitat fragmentation. Those who study the mechanisms of the biodiversity crisis like to say that theirs is the science of scarcity and diversity, but what they are contributing to human knowledge is a recognition that natural systems at all scales are in constant states of flux. Even Koma Kulshan will some day slip into the sea.

This position is at odds with the view of nature that we have inherited from our Euro-American ancestors, a combination of the mechanistic Newtonian-Cartesian clock combined with theories of balance and harmony borrowed from nineteenth-century Romanticism. Behind the traditional view of Douglas-fir forests achieving maturity in old growth lies the notion of orderly equilibrium. But ecological theory is changing. The balance-of-nature concept "makes nice poetry, but it's not such great science," remarks Steward Pickett.[102] What biologists see instead is continuous change. The shifting-mosaic, steady-state pattern of disturbance has been the norm for as far back as ecologists can probe. Today's plant and animal communities are not fixed—most date back less than 10,000 years, as far as the last major shift in world climate.[103] Even over periods "as long as anyone would claim for 'ecological-time,' there has never been an interval when temperature was in a steady-state," reports paleoecologist Margaret Davis.[104]

Biosphere patterns shift constantly, and landscapes, ecosystems, and species respond in kind. As climate changes, we can expect all manner of rearrangement in species and ecosystem patterns and processes—expect, but not predict precisely. Historian Donald Worster sums up the emerging role that uncertainty plays in evolving ecological images of nature:

> Nature, many have begun to believe, is fundamentally erratic, discontinuous, and unpredictable. It is full of seemingly random events that elude our models of how things are supposed to work. As a result, the unexpected keeps hitting us in the face. Clouds collect and disperse, rain falls or doesn't fall, disregarding our careful weather predictions, and we cannot explain why. A man's heart beats regu-

larly year after year, then abruptly begins to skip a beat now and then. Each little snowflake falling out of the sky turns out to be completely unlike any other. If the ultimate test of any body of scientific knowledge is its ability to predict events, then all the sciences . . .—physics, chemistry, climatology, economics, ecology—fail the test regularly. They all have been announcing laws, designing models, predicting what an individual atom or person is supposed to do; and now, increasingly, they are beginning to confess that the world never quite behaves the way it is supposed to do.[105]

This is not just a battle between theoretical schools. The lessons emerging from the biology of thinking like a mountain are as profound as determining the prospects of survival for unknown numbers of living beings. Homo sapiens is not automatically exempt from this group.

If conservation biologists were pressed, they might distill their theories into this advice: Think big, think connected, think whole. In the face of the largest human experiment in history, the loading of Earth's atmosphere with excess carbon, this counsel may be difficult to follow. Climate change on such a broad scale over such a remarkably short time will so affect biodiversity that not only animals and plants will need to migrate but "preserves themselves may need to move."[106] Designing biological corridors to facilitate this scale of landscape change boggles the mind; yet some biologists have already published papers predicting the need for continentwide, north-south connections between protected areas.[107] Elevational corridors may also be necessary. As temperatures warm and less moisture is available at low elevations, many plants may have to climb toward favorable conditions. Contrast this developing consensus among biologists—to think whole—with the attitude of the head of the U.S. delegation to the Second World Climate Conference in November 1990. When pressed by the majority of delegates to support specific restrictions on the output of greenhouse gases, Dr. John Krausse held to the Bush administration's party line: "We just don't believe in targets."[108]

The practical problems of global greenhouse bargaining are linked with protecting a sustainable system of parks and reserves and managing for viable populations of grizzly bears. All are manifestations of the same problem at different scales of the biodiversity hierarchy. If nature is uncertain, if there is no balance but only a "discordant harmony," as Daniel Botkin calls it, then what are people bound to do? In a world where perturbations are the rule, where chaos reigns, what purchase do we have, what goals may we set?

Botkin points out that "there are ranges within which life can persist, and changes that living systems must undergo in order to persist. . . . Those changes that are necessary to the continuation of life we must allow to occur. . . ."[109] Conservation biology is teaching us about the limits of living for Homo sapiens as well as endangered species and ecosystems. Wildlife biologist Malcolm Hunter clarifies the issue: "Today's distribution of species and communities is . . . one frame of a movie recording continuously changing distributions and associations of (organisms)."[110] Koma Kulshan's perspective tells us that we must maintain healthy populations, expand existing reserves, and restrain our drive to burn every remnant of the Carboniferous period so that the other frames of the movie are not edited into oblivion. If we don't, there will likely be a giant splicing of biodiversity whose ramifications are uncertain.

In an untidy world, humans can no longer expect uniform solutions to complex problems. And, while whales are lovable and ancient forests inspiring, we must also learn to recognize that *all* living beings depend upon the systemic functioning of their native habitats, however unpredictable that may be. Though many conservation-biology theories remain untested, the maturation of the field may have less to do with developing formal rules and principles than with helping people to learn what already appears obvious: Humans cannot for long impose limited notions of order on a living world that, by its very nature, will not be pinned down. There are no grounds for restricting our interactions with nature to any one level of biodiversity. The experience of Koma Kulshan knits levels together. Once we pay attention to this, the only word left to balance uncertainty is respect.

CHAPTER THREE

GHOST BEARS

> *. . . Bears precede our perception of them, they came before national forests, before the word* forest*—black, shaggy beings emergent from millions of forested years without benefit of manufacture or legislation.*
>
> DAVID RAINS WALLACE,
> *The Klamath Knot*

It is August of a dry summer after a record wet spring and I am on a high heather ridge just below timberline. Here in the Glacier Peak Wilderness, the backcountry heartland of the Greater North Cascades ecosystem, it is the onset of berry season. For grizzly bears, late summer through fall is a critical period. They must consume vast quantities of carbohydrates to gain weight before winter denning. This feeding, called hyperphagia, may involve an intake of 20,000 calories per day. In the North Cascades, this means berries—huckleberries, blueberries, serviceberries, raspberries, elderberries, chokecherries, mountain-ash, honeysuckle, kinnickinnick, and more. Bears supplement this fruit diet with tubers, roots, insects, grubs, rodents, and whatever other small animals they can claim. As the world's biggest omnivore, the grizzly displays a taste for diversity.

It is huckleberry time in the high Cascades. I work north along the ridge, angling toward a green avalanche chute lush with shrubs and herbs. The wet smell of subalpine meadow rises up around me as I swim through Sitka valerian and false hellebore. Three of the most-used grizzly habitat types in the Greater North Cascades make up a mosaic of the land I travel: alpine ridge, berry shrubfield, and talus. Below me, I scan subalpine mountain hemlock forest mixed with flat meadow openings. This is ripe late-summer bear habitat.

I am not seeking grizzlies. Instead, I am paying my respects to the end of the growing season, positioning myself for encounter without expectation. I am exploring heightened possibility and perception.

One learns to look with care in the mountain world. Narrow vision blinds. Best to focus, instead, on peripheries, movements. A rockslide may hide a pika gathering hay for winter. A tiger lily dissolves into a swallowtail butterfly. The avalanche track I am now walking on has turned from a green jungle into intertwined willows, alders, and elderberries. It is so choked with vegetation that if a grizzly bear were nearby, I would only notice it by movement in the crown of shrubs or the crack of breaking branches. Flies buzz around my head. I can peer into the chute for a distance of no more than fifty feet. I hear only the flies and falling water rising from the basin below. Plant odors waft up from the tangle, but I cannot distinguish the scent of an elderberry from an aster.

Grizzly bears can smell a human about 5,000 feet upwind, can detect carrion as far away as seven miles. At a distance of 1,000 feet they can hear human conversation, and they can run as fast as thirty miles per hour. I have never seen or heard a grizzly in the wild. The closest I ever (know I) came to encountering one was in the south fork of the Snake River, hard by the southern boundary of Yellowstone National Park. On a maintained trail I crossed a fresh track in mud no more than a few hours old. Kneeling down, I felt the hair on the back of my neck rise with the realization that two beings had almost met in a vast, wild place. I knew how uncertain the likelihood of meeting *Ursus arctos* was in country marked by long swirling rivers, dense forests, and miles of upthrust highlands, she merely ambling across the trail and I following someone else's idea of passage.

There are now fewer than 200 grizzlies left in Greater Yellowstone, and there are none in this North Cascades avalanche track—nor anywhere else on the ridge within my view. Yet this is high-quality bear habitat in the wildest mountains in the Pacific Northwest.

Where are the bears?

To many, the grizzly bear has lived in the mountains forever. Some modern-day hikers, timber-company executives, and not a few federal bureaucrats maintain, however, that the bear is gone for good; they want to release trails, forests, and multiple-use land management from environmental restrictions. Timeless presence or threatened species? These two views result from a potent mix of different time frames and interests. The politics of endangered species and ecosystems has to do with deep, conflicting myths that feed our cultural roots.

The grizzly embodies all the problems and promise of the biodiversity crisis in the North Cascades. The bear has been protected by the Endangered Species Act since 1975, yet until recently no study had been initiated to determine the status of the species in the region. There are certainly grizzlies in the North Cascades, but there have been no definitive sightings for years. The bear has, so far, slipped through the net of scientific inquiry we have thrown over the mountains. Given our modern fear of predators more powerful than we, the limits of our knowledge translate into uncertainty. The grizzly bear is a ghost in the Greater North Cascades, off trail, beyond the campfire's light, living in the hills unseen, playing at the edge of dreams to some, nightmares to others. In the deep past, humans developed a complex relationship with the bear based on respect. Today, the ghost bear fires the imagination of people who want to practice ecological restoration of the North Cascades; the animal is the classic umbrella species of conservation biology. It is also the preeminent indicator of wildness. For those who want to use wilderness resources, the grizzly is an obstacle that must be removed.

I have never approached a grizzly bear, but what I have discovered is this: The largest, most powerful animal living in the Greater North Cascades ecosystem is a modern apparition that, some day, may well be nothing more than a dweller in the shadowlands of memory.

Grizzly Bears in North America

When it comes to grizzly bears, the issue of conflicting time frames is easy enough to resolve. I know that the genus *Ursus* traveled over the Bering land bridge to North America from Asia about 250,000 years ago and has been evolving into distinct subspecies ever since.[1] As late as 1800 A.D. there were roughly 100,000 grizzlies from Ontario to the Pacific Coast and south into Mexico. Throughout the North Cascades, with the possible exception of the lowlands on the west slope, there existed a healthy population.

The arrival of Euro-Americans in the North Cascades, as elsewhere, precipitated a dramatic decline in grizzly numbers. The decimation occurred in four overlapping waves. In the early to mid 1800s, trappers killed many bears. In only five years' time (1846–1851) Paul Sullivan documented 423 hides taken from the North Cascades.[2] Miners, pursuing gold and other precious metals, followed the trappers into the mountains. Though no specific records of animal casualties remain, the count must have been high because of indiscriminate shooting and habitat encroachment as the prospectors scoured creeks and ridges for strikes.

Ranchers running large herds of sheep and cattle into the eastside high-country meadows next came into conflict with the bear. The heyday of livestock grazing in the North Cascades lasted from the 1850s into the early years of this century. And, by the 1870s, logging was spreading up the lowland valleys on both sides of the range. Timber management has since become the dominant human activity in the North Cascades.

Is the grizzly bear endangered? The answer to this question depends partly on where you draw your timeline. From the standpoint of the last 150 years, the grizzly has experienced catastrophic decline. Hard evidence suggests that there is not enough living space south of Canada for both bears and humans: 99 percent of the grizzlies have been destroyed. There are now less than 1,000 left, confined to six separate ecosystems in the wildest parts of the western United States: the Greater Yellowstone, northern Continental Divide, Cabinet-Yaak, Selkirk, Bitterroot, and Greater North Cascades ecosystems (see figure 4).[3] The great worldwide refuges of wild *Ursus arctos* are in Canada and Alaska and in the former Soviet Union, with 20,000 and 100,000 animals, respectively. Of the six remaining grizzly-occupied ecosystems in the continental United States only two, Greater Yellowstone and the northern Continental Divide, have reliable population estimates.

In the North Cascades, those who wish to see the last grizzly gone

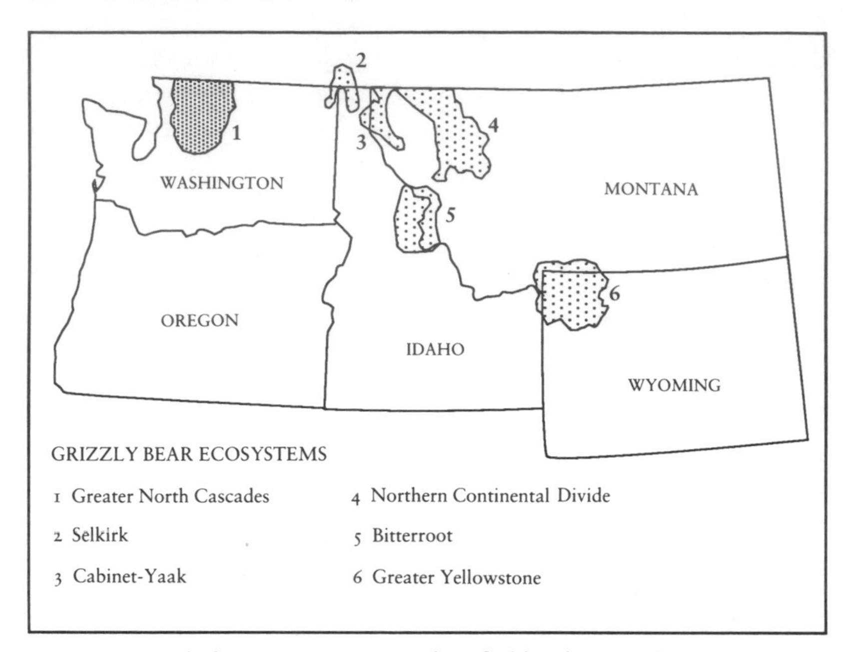

FIGURE 4 Grizzly bear ecosystems as identified by the Grizzly Bear Recovery Plan.

Source: U.S. Department of Interior, Grizzly Bear Recovery Plan (Washington, D.C., 1982), 195.

have narrowed their focus to the past six years. Since 1986, Washington State Department of Wildlife biologist Jon Almack has been searching for the bear as he evaluates grizzly habitat suitability. Almack's study is part of the work of the Interagency Grizzly Bear Committee (IGBC) which, since 1983, has coordinated federal, state, and private research and management of grizzlies. The IGBC is responsible for overseeing recovery efforts on the animal's behalf. If Almack can find sufficient high-quality habitat in the Greater North Cascades, the decision can be made to initiate full-scale recovery efforts.

Almack has found plenty of bear habitat and signs, but he has never found a live bear. His search has sparked considerable controversy in the region: Hikers are afraid of bears; loggers dread another northern spotted owl issue; land managers are apprehensive about yet one more claim on their decision-making authority.

For Jon Almack, a study funded only for five years is too little time to become familiar with 6 million acres of wildlands, let alone find one of what is certainly no more than a handful of resident bears. A Forest Service supervisor sees five years of coming up empty-handed, with but a few photographs, some plaster casts of tracks, and several arguable sightings, as proof enough that proposed roads, timber sales, and ski resorts should proceed.

The present-day North Cascade range is relatively young, less than 10 million years old. *Ursidae*, the bear family, has three times journeyed to North America over the last 2 million years. In evolutionary terms, that is a long time ago. David Rains Wallace has suggested that "bears came before the *word* forest. . . ."[4] Certainly the grizzly preceded what we now recognize as the ecosystems of the North Cascades. Paleobotanical studies carried out by Linda Brubaker of the University of Washington show that Douglas-fir forests only began to dominate the Pacific Northwest about 10,000 years ago.[5] Old-growth forests as we know them have only existed for 5,000 to 7,000 years.

How do humans fit in with the evolutionary journies of bears, forests, and landscapes? Are there lessons to be learned from our nonhuman kin? Part of the answer lies in an old story from long ago.

The Bear Mother Myth

Once upon a time there was a girl out picking berries. She was a young girl, strong, and used to having her own way. One afternoon, she stayed late picking berries and everyone else went home. She spilled her basket and was gathering them up. A tall, handsome man appeared. He helped her and said, "Let's keep picking till dark. I know good places and will take you home." They picked together in the gathering dusk. Then it was dark. They built a fire, roasted meat, and ate. They slept by the fire. The next day they began picking berries again but the girl thought about home. "I want to go home, now," she said. The man slapped her on the head and circled her head with his finger the way the sun circles the Earth. Then she lost track of time and they traveled and picked berries for a long while.

He was just like a human, but she thought he might be a bear. They traveled a long while. It was fall now. It was turning cold. She knew that he was a grizzly bear. "It's time to dig a winter lodge," he said. "Get some brush for our bed." He began to dig into the side of the mountain. She collected fir boughs but she left a sign for her brothers, too. "That brush is no good," he said, "and you have left a mark to guide your brothers here. We must go." They traveled up the valley to a new place. He began to dig again and she collected brush. This time, she brought the proper bedding. But she also left a sign. She knew that her brothers would hunt the valley in the spring and find their den.

During winter, she bore two babies, a girl and a boy, just when bears have their cubs. The winter was very long. One time she woke to hear her husband singing. He was a shaman. "Before the snow is gone your brothers will come for me—I will fight them!" he said. "No! No!" she replied. "They are your relations now, you cannot kill them." "All right, I will let them come," he said. But later, he was singing again. "If they come, if they kill me because I cannot fight them, get from them my skull and tail. Burn my head and tail in a fire and sing my song till nothing but ashes remains." Then they slept. Spring was coming.

Now he was singing again. Now the dogs were barking and the brothers were in the valley hunting. "They are coming! I will fight them to the death!" "No, you must not! Who will take care of the children if they lose both their uncles? You must let them come." "I am taking down my knives," he replied. He was a very large grizzly bear. "I am going now." He left the den and it was silent for a long time.

The woman left the den. She came out into the spring light. Her brothers had killed the grizzly bear. They found her on the mountainside. "You have killed your brother-in-law," she said. "Save the head and tail for me."

She went home to her family after being gone a year. They hardly recognized her. That night, she built a great fire and burned the grizzly's skull and tail and sang his song. She could not stay in her mother's house. She built a separate camp nearby for herself and her children. In the fall as the air turned cold, she finally came into her mother's lodge. It was not easy for her to do this.

Next spring, her brothers killed a female grizzly bear with two cubs. They wanted their sister and her children to wear the hides and pretend they were bears. "No, I cannot do that or I'll turn into a bear forever," she said. But her brothers would not listen. They pestered her. They wanted her to put on the hide. Then they snuck up from behind and threw the hides over them. She turned into a big grizzly bear and walked on all fours. Then she had to fight them. She killed her family, sparing only her youngest brother. Tears were streaming down her face. Then she took her children, her cubs, and went back into the mountains, far away.[6]

The story of the woman who married a bear, or the Bear Mother, reveals just how deeply connected bears and humans are. The tale, or more properly myth, "may be the most persistent and widely told tale ever devised to entertain and educate."[7] It explores the kinship between people and bears, of how a woman crosses the boundary between species through marriage and childbirth, how her bear-husband dies as a result, and how she serves as a messenger between two worlds. The tale tells of identity, distinction between the species, appropriate ceremonies to honor the dead bear. It derives from cultures in the northern countries of both East and West, as far south as Iran in Asia and South America in the western hemisphere. In the North Cascades, anthropological research shows that the Upper Skagit, Methow, and Thompson tribes all honored the grizzly bear.[8] After a successful hunt the bear's skull was placed on a pole in the woods to assure perpetuation of the species. The skull is the seat of the soul, and a grizzly's death, if properly respected, is momentous. Death mirrors life: the animal's hibernation is followed by rebirth in the spring. Religious historians Ake Hultkrantz and Iver Paulson have noted that the careful treatment of grizzly skulls is a widespread practice and signifies seasonal change, rites of passage, and renewal.[9]

The Bear Mother myth gave birth to a myth about her two sons. Paul Shepard and Barry Sanders point out the subtleties that separate the two stories: While the first portrays humans receiving the bear's message as a gift, the second shows the Bear Sons seizing what the world has to offer through their own strength and skill.[10] Human cleverness is on the rise; the power of the bear has begun to recede.

What has been sung in myth throughout the ages is today mirrored in the fate of the grizzly of the North Cascades.

Grizzlies in the North Cascades

From a distance, the world is blue and green and Mount Rainier is a hulking mirage on the southwest skyline. I am perched on a ridge just off the Cascade Crest, somewhere above the avalanche chute. Endless mountains rise up in all directions and the sky is as clear as snowmelt from an alpine tarn. Rocks are buzzing in the heat. Flowers wave bright wands of color. There are goat tracks pressed into the scree, melted into hoof-shaped cups of snow.

By these signs, and others, I know I am alive. Breath comes short from the climb out of the gully. The unobstructed view guides my vision from horizon to horizon. Koma Kulshan, hiding behind the left shoulder of Glacier Peak, stares me in the eye. In the continental United States there are few places this wild outside of the northern Rockies, and none in the Northwest that offer grizzly bears the habitat they need in such seemingly extravagant abundance.

As a result of the pioneering work of the Craighead brothers, bear biologists recognize space as the first of seven key habitat characteristics that define the needs of grizzly bears.[11] The North Cascades Grizzly Bear Working Group, an ad hoc interagency committee of which Jon Almack is a member, has identified the outer boundary of the North Cascades grizzly bear ecosystem as encompassing 5.7 million acres of federal land and another 300,000 of state and private holdings.[12] This immense area is marked by the international border to the north, the Puget Sound lowlands to the west, the Interstate 90 corridor to the south, and the eastern limit of national forest land near the Okanogan and Columbia River valleys. The bulk of this mountainous area is within the administrative boundaries of the Wenatchee, Okanogan, and Mount Baker–Snoqualmie national forests and the North Cascades National Park complex. Within this greater ecosystem, Almack has delineated a protected "ecosystem core" of about 2.5 million acres comprised of the national park, contiguous Forest Service wilderness lands, and adjacent protected areas in British Columbia.

As far as grizzly bears are concerned, the size of this ecosystem core is deceptive for four reasons. First, about 19 percent of the area is glaciers, snowfields, bare rock, lakes, and habitat similarly unsuitable for bears.[13] Second, the Canadian lands were not included in Almack's final study because of "political boundaries and administration." Though there were ten sightings reported between 1972 and 1985 just north of the border, no one knows the status of the bear in British Columbia.[14] Third, Almack based his grizzly population estimates on the core area, but grizzlies do not limit themselves to ecosystem cores, the North Cascades grizzly bear ecosystem, or international boundaries. Finally, the concept of an ecosystem core diverts attention from the matrix of multiple-use lands within which the core is embedded. It is on these unprotected lands that the fight to protect the bear will be won or lost.

There is always politics at play in our efforts to define the limits of our use of nature. The world, no matter what our intent, is not easily captured by any one human theory, value, or standard. Pressure is heightened when we grapple with the vital needs of an endangered animal that depends on vast areas of wildlands. And we deceive ourselves when we focus on managing the grizzly as an antidote to our own desire to develop the land that sustains the bear.

How many grizzlies can be sustained within the ecosystem core of the Greater North Cascades? Jon Almack's estimate of 85 to 130 bears illustrates the strain between two species' demands for space as well as the crucial need for integrated ecosystem management that goes well beyond the boundaries of the North Cascades. Since there is little specific data on the size of the North Cascades grizzly population or its demographics and habitat use, Almack's estimate is based on data and models imported from the Greater Yellowstone ecosystem, where the bear has been studied for over twenty years. Almack's calculations assume a high degree of similarity between the Yellowstone and the North Cascades in average home range for adult female grizzlies, nonoverlapping home ranges, percentage of female bears in the total population, and corrections for the fact that not all suitable habitat is occupied at any given time. To a bear lover, 130 grizzlies in the North Cascades sounds like good news. But this number represents only what the ecosystem core may be able to support at full capacity.

To a conservation biologist, 130 grizzly bears is a dangerously low population. Theory, as recounted earlier, suggests that to sustain a viable population of large mammals over the long term requires an effective population of at least 500 animals. And Ne sizes are almost always lower than the census numbers that Almack's estimate is based upon. The most authoritative estimate for grizzlies published so far concludes that 1,670 to 2,000 censused bears would be necessary to sustain a single isolated viable population.[15] One hundred thirty is not sufficient to sustain the grizzly in the Greater North Cascades.

Jon Almack's concept of the ecosystem core is not sufficient either. What about the lands in the North Cascades grizzly bear ecosystem outside of this core? Almack does not factor these lands into his population estimate, and so it is not indicative of the region's full habitat capacity for the grizzly. What 130 bears does represent, as pointed out by Mitch Friedman of the Greater Ecosystem Alliance, is "the number of bears that could be expected to live on lands *already* protected within the Greater North Cascades."[16] Friedman provides the missing link between the ecosystem-core estimate and a full-scale greater-ecosystem perspective. The question then becomes how many grizzlies, given Almack's assumptions and calculations, might the *entire* 6 million–acre North Cascades grizzly bear ecosystem support?

The answer is between 246 and 377 bears. A North Cascades population of 377 grizzlies would be encouraging, but it is still a far cry from a long-term viable population. In fact, the space needed by 1,670 censused bears renders even the largest Greater North Cascades boundary inadequate. Theoretically, that many bears would need over 40,000 square miles of suitable habitat, an area equal to about 56 percent of the state of Washington. The most visionary proposal in favor of the grizzly to date would only protect habitat for 352 to 554 bears.[17]

What these numbers illustrate is how little we really know about *Ursus arctos*'s needs, and, at a deeper level, how fragmented our modern concept of nature has become. Extrapolating field data from one place and applying it to another is always risky business: The Greater North Cascades is not Greater Yellowstone. In the North Cascades adult female grizzly bears may have smaller home ranges. If this is true, a viable population might depend on less area. Conversely, North Cascades female bears may need more space than those in Yellowstone. No one knows. Sullivan's historical data on the number of grizzlies killed in the nineteenth century suggests that hundreds of bears lived in the North Cascades. But his observations, while tantalizing, cannot prove anything about population size, home range, or viability.

The important thing to remember about viable population theory is that it is just that—theory. Little of it has been tested using empirical data. It could be that a viable census population size for grizzlies ranges much lower than 1,670 to 2,000 animals. Or it could be higher. Without a working knowledge of grizzly population demographics based on field observations, viable-population theory can never be tested. This data must be gathered by tracking radio-collared bears. Consider that after his five-year search in the North Cascades, Jon Almack hasn't seen or photographed a single grizzly, let alone radio-collared one.

Whatever the limitations of our data and theories, we do know that grizzly bears need a lot of room to roam. Based on Greater Yellowstone data, Jon Almack calculates that an average bear uses a home range of about 100 square miles.[18] With the help of the bear, we are coming to realize that the wild habitat we have protected in parks and reserves is not enough to serve the grizzly and many other wide-ranging animals. When we regard ecosystem cores as specific units of land separate from the North Cascades grizzly bear ecosystem, we are prey to the cookie-cutter consciousness that sees nature as a set of distinct political entities instead of a unified whole. A grizzly bear ecosystem cut off from the landscape of mountains, forests, and sagebrush plains that we recognize as the Greater North Cascades would not serve the bear well. Even an international Greater North Cascades, going far beyond present political possibilities, would float disconnected in a regional matrix of developed lands. Mitch Friedman and other conservation biologists talk of maintaining several interbreeding subpopulations of *Ursus* connected by habitat corridors. For North Cascades animals, this would mean fostering habitat connections (currently severed) between southern British Columbia and the Selkirk grizzly bear ecosystem 120 miles to the east. The number of grizzlies in any one of these ecosystems does not meet viable-population criteria, but in a reestablished gene flow there lies hope for the bear.

This hope is based on what would be unprecedented behavior on the part of humans in support of *Ursus arctos*. Maybe the grizzly is too big for us to understand or allow to survive. Certainly we will be tested by our upcoming actions for and against the bear. But the grizzly has always assayed our willingness to come to terms with, in the poet Gary Snyder's words, "matters of manners between species."

Gracious accommodation of the grizzlies' need for space is but the first matter of manners essential for quality habitat. Isolation—available space considered along with amount of human activity—is also important. North Cascades grizzlies live in a region where human development is not yet far advanced. Only three major highways bisect an area of close to 9 million acres. Towns are limited to the periphery. Ironically, lack of human presence, which so far has allowed the grizzly to persist, albeit in small numbers, serves as a powerful inducement to those who see the world as a cornucopia to be raided. On the park and wilderness lands within the North Cascades grizzly bear ecosystem, this threat poses no problem. On the multiple-use lands that constitute roughly 70 percent of the greater ecosystem, developments are on the drawing board.

Three factors play major roles here: road building (primarily for logging), livestock grazing, and to a lesser degree, loss of low-elevation habitat.

Roads are a major threat to grizzly bears. They provide human access, which increases the potential for conflict between bears and people. Grizzlies may benefit from the rich food sources that appear in clearcuts after logging, but this advantage is offset by hunting, poaching, and loss of isolation. Numerous studies bear this out. In the longest-running research project on the effects on grizzlies of logging, mining exploration, and other development, Bruce McClellan and David Shackleton found hunting and poaching that resulted from road access to be the number one cause of death in the animals they were working with.[19] In Montana, Arnie Dood and his coworkers discovered that 48 percent of all known grizzly mortalities occurred within one mile of roads.[20]

Many biologists have also noted that grizzly bears tend to decrease their use of suitable habitat after roads are opened up nearby. David

Mattson and his associates found that grizzlies do not usually wander within 1,600 feet of roads during spring and summer and they stay even farther away in the fall.[21] In the coastal forests of British Columbia, biologists have noted a 78 percent decline in habitat use within 500 feet of active logging roads.[22] McClellan and Shackleton, working in the northern Continental Divide ecosystem in British Columbia, northwest of Glacier National Park, found a similar decrease. And Wayne Kasworm and Tim Manley discovered that bears in the Cabinet-Yaak ecosystem in northwest Montana avoided habitat within 1,300 feet of roads.[23] These last two biologists point toward the more subtle effects of roads on grizzlies: "Grizzly bear avoidance of high-quality habitat near roads and trails may lessen the opportunity for individuals to obtain food and increase intraspecific competition by further forcing the bears into limited remote habitat."[24] Bruce McClellan disagrees: "If logging operations came and took the timber properly, and shut the roads, there would be no impact on the bears."[25] McClellan speaks from twelve years of experience studying the effects of multiple-use activities on grizzlies. Nevertheless, it is unusual to hear a scientist speak in such unqualified terms; no one would suggest that our knowledge of the grizzly's response to development is complete.

Disagreements aside, all bear biologists stress the need to reduce conflict through an aggressive road-management program. In bear habitat, closing roads would reinstate the isolation the grizzly requires and protect it from poaching and harassment. Activist biologists Mitch Friedman and Tony Povilitis suggest that on federal lands in the Greater North Cascades, all roads should be closed within key grizzly bear habitat. Outside key habitat, road density should be kept below one linear mile/square mile. They also recommend a road-building moratorium until a bear-recovery plan can be implemented. There is little scientific data to support the first two of these claims. We don't know where key grizzly habitats are in the North Cascades, and road-density studies are lacking. But this is an example of where scientific uncertainty should not stand in the way of action to protect the bear. What Friedman and Povilitis are reacting to is the road-building plans of the Forest Service on the national forests that comprise the bulk of the Greater North Cascades.

Of the three national forests within the region, the eastside Wenatchee and Okanogan national forests have the choicest grizzly habitat. They are also subject to the most potentially damaging road-building plans. The Wenatchee forest plan proposes 1,450 miles of new construction over the next twenty years.[26] Half of these roads would be built in currently unroaded areas of forest. The Okanogan forest plan allows for 330 miles of new roads over ten years, with 55 percent of unprotected forest lands affected.[27] The westside Mount Baker–Snoqualmie forest plan outlines 130 miles of new road over ten years, with over a half the mileage in roadless areas.[28] None of these plans discusses the potential effect of such development on North Cascades grizzly bears. There is no attempt to evaluate issues based on scientific literature. Nor is there any research proposed in the plans that would help resolve such questions. With a grand total of almost 2,000 miles of new roads proposed across the Greater North Cascades on federal lands alone, Friedman and Povilitis are justified in calling for a road construction moratorium until the effects on the remnant grizzly population can be ascertained.

Livestock grazing, though less damaging than roads, has also reduced isolation for North Cascades grizzly bears. But both the Wenatchee and Okanogan forest plans propose to increase grazing above recent levels. Research conducted in Greater Yellowstone has found that livestock-related conflicts may have been responsible for up to 10 percent of grizzly mortalities between 1960 and 1970.[29] In the North Cascades, a 10 percent reduction over ten years could mean extinction. There is no discussion in the forest plans of the potential effects of increased grazing levels on grizzlies. The issue of extinction aside, the question is whether current and proposed grazing levels prevent the few remaining grizzlies from occupying habitat that may be essential for population recovery.

Loss of isolation due to human usurpation of low-elevation habitat for towns, farms, and ranchland has pushed the grizzly farther back into the mountains. This loss, in itself, is probably not a limiting factor for the bear in the North Cascades. The common Western perception is that population growth and more use of resources are good and necessary. As we run out of room in the lowlands, we set our sights on the upper reaches of the valleys. The grizzly bear can only be pushed so far.

The proposed Early Winters ski resort on the Okanogan National Forest is a case in point. If built, the ski area would attract thousands of new visitors to an area where grizzly sign has been sighted. What negative effects could wintertime skiing, when bears are safely hibernating, have on the grizzly? According to an environmental impact statement (EIS), the project would reduce the Okanogan Valley deer herd, at 18,000 animals one of the largest in the state, by 50 percent.[30] Those grizzlies that may depend on the herd for winterkill carrion or growing-season prey would find this food source reduced by half. But our concern for the grizzly bear's needs lags behind our desire for another destination resort. The Early Winters EIS does not address impacts on grizzlies.

The proposed ski development illustrates the importance of space and isolation for grizzly bears. Sanitation, another critical habitat factor, is not now a problem in the North Cascades. It is well known that grizzlies can become habituated to human food and that this often leads to problems for both bears and people.[31] But Jon Almack has noted little response by either the federal agencies or the general public to his efforts to encourage appropriate food use and disposal. In the back-country, few hikers and agency employees use sanitation safety tips that are routinely employed elsewhere in occupied grizzly habitat. Front-country campground dumpsters and garbage cans are allowed to overflow regularly. There is no move afoot to bearproof park and forest campgrounds. (An outstanding exception to this is the new car campground at Hozomeen on Ross Lake in the North Cascades National Park complex.) Managers and citizens respond to Almack by pointing out that there is no habituation problem in the North Cascades; certainly campers have not encountered the bears that biologists conducting an intensive survey have found so elusive. It will take only one negative encounter between grizzly and human, Almack replies, to jeopardize the bear's future in the North Cascades.

Judging from public reaction to the North Cascades grizzly, Jon Almack may well be right. In 1989, in public testimony not heard before from users of other grizzly bear ecosystems, a significant number of hikers joined with developers and other groups traditionally allied against the bear to denounce federal recovery efforts. Each year 800,000 hikers use the Cascades; they form a powerful lobby. According to journalist B. J. Williams, "a lot of Washingtonians have developed a recreational style based on the idea that you can have wilderness without significant danger or even inconvenience."[32] It is no matter to these hikers that they are fifty times more likely to drown in their bathtub than be killed by a grizzly bear. Or that the North Cascades is the grizzly's home. Public attitudes illustrate the uphill battle the bear faces in recovering its territory. They also prove Jon Almack's point: Sanitation concerns must be addressed if the grizzly bear is to succeed in the North Cascades.

Bears also need undisturbed places to den for the winter. Grizzlies move into their winter homes in November and emerge between March and May. They prefer excavated chambers on subalpine slopes with good drainage and deep snow accumulation. Jon Almack has found no lack of suitable den sites in the Greater North Cascades.[33]

Two other key habitat factors, vegetation type and food availability, are still poorly understood. Almack has had to extrapolate data from other, better known grizzly ecosystems and apply it to the North Cascades. For bears that are not radio-collared, preferred habitat and diet are difficult to fathom. What we do know is that eastside vegetation is more similar to what the bear prefers in the Rockies than the plant communities found on the westside. Westside vegetation includes, however, a rich supply of berry-producing shrubs. Bears depend on vegetation when it is at or near ripeness and when it is easily digestible. Many specific grizzly food plants grow on both sides of the range. The infamous North Cascades weather, unpredictable and at times unseasonable enough to turn July into January, may, in cooler years, affect food availability. Almack suggests that this could be a limiting factor for North Cascades animals.

Safety from human activity is the last key habitat factor, implicated in all the others. A den site must be located away from disturbances, and food supply must not be compromised by ski areas. Almack lists timber harvesting, grazing, recreation, and land development as the major safety issues in the North Cascades. If these activities occur near or within bear habitat, they "must be tailored to allow for the coexistence of grizzly bears."[34] This requires coordinated management among many federal, state, and private groups. For example, if the ecosystem core is to serve as the gene pool for an ecosystemwide recovery effort, then surrounding multiple-use lands must be managed to maintain if not enhance suitable habitat. Grizzlies cannot survive in the ecosystem core alone.

Grizzly bears depend on us. There is little chance of their surviving in any continental U.S. subpopulation unless humans are willing to give suitable habitat back. Reconnecting subpopulations with biological corridors would decrease the genetic, demographic, and environmental uncertainty threatening the species.

But before we reconnect landscapes, we must reforge our bond with the bear. There is no escaping the fact that fear is what many people feel when a grizzly ambles into view. As yet, there is no clear consensus *for* the bear in the North Cascades. What appears to be a long-standing dispute between two species over living space, however, shrinks in the context of the full sweep of history and myth. One of the lessons of the Bear Mother myth is that sustainability for bears and humans is somehow linked. Mountain people of the past knew this, that they were dependent on the bear. Perhaps it is time for a twentieth-century ritual gesture: instead of burning bear skulls as our ancestors did, we might weave the grizzlies' fragmented habitat back together.

Grizzly Bear Recovery

I am tired, hungry, and it is late afternoon at the end of a long day of nosing into avalanche chutes and following game trails. Again, good bear habitat, no sign. A single raven flaps overhead, wheels around, is gone. I am ready to be restored.

Are we ready to restore the grizzly bear to the Greater North Cascades, allow the grizzly to recover its former range?

On March 3, 1990, the Humane Society of the United States and the Greater Ecosystem Alliance formally petitioned the U.S. Fish and Wildlife Service under the Endangered Species Act to upgrade the North Cascades grizzly population from threatened to endangered. In August 1991 the agency rejected the petition, stating that the listing was "warranted but precluded due to funding limitations and other priorities."

Tony Povilitis and Mitch Friedman, the scientists behind the petition, are not conventional biologists. This is clear by the very fact that they have taken a political position. The points of the petition itself were by no means exceptional. The appeal was based on the facts of life for the grizzly in the Greater North Cascades: an extremely small population, proposed increases in road building, grazing, and land development, no information on the cumulative effects of these activities, and lack of affirmative action by the Forest Service, Park Service, and Fish and Wildlife Service on behalf of the bear. The first of these charges is easy enough to document. The last point, until recently, was more difficult to prove.

The IGBC may have shown its hand with the publication of the revised Grizzly Bear Recovery Plan. Released in October 1990, this draft plan plots the future of the bear in the six occupied grizzly ecosystems in the continental United States.[35]

There is very little from conservation biology in the draft plan. It treats each subpopulation as an isolated ecosystem and assigns a recovery figure well below what biologist Fred Allendorf's work has shown to be "necessary to maintain evolutionarily significant quantities of genetic variation."[36] There is no discussion of the loss of linkages between subpopulations, even though Greater Yellowstone bears have been isolated for at least sixty years. Nor is reconnecting the islandlike ecosystems through biological corridors considered. The federal strategy is based instead on "population augmentation," where bears would be moved by managers from one region to another. The four subpopulations that currently share some degree of genetic interchange with Canadian bears, including animals from the North Cascades ecosystem, are assumed to be permanent beneficiaries of this arrangement, even though protective legislation is nonexistent across the border and the pace of logging in British Columbia can only be described as frenzied. The plan appears to freeze Canadian grizzly populations in time. But expert bear biologist Dr. Stephen Herrero of the University of Alberta warns, "Don't count on Alberta saving grizzly bears for America. It's likely to be the other way around."[37]

The recovery plan does not deal with recovery at all. It does not "seriously address the implications of isolation and the effects of habitat dynamics" on the six subpopulations. If grizzlies are to recover, they must increase in number toward what conservation biologists consider a viable population, and to do this they must have more habitat instead of being trapped in so many ecosystem "zoos." Recovery targets for the northern Continental Divide ecosystem around Glacier National Park tell the story best. The 1990 proposed target in this ecosystem, 440 bears, is the same as the *lowest* estimate for the population in 1980.[38] Seventeen years ago, a number such as this was grounds for listing the grizzly under the Endangered Species Act. Today, according to the IGBC, it signifies success.

The North Cascades subpopulation is faring no better. The target of seventy to ninety bears under the existing recovery plan is discarded by the new proposal.[39] In its place is a "further evaluation" category where "areas will be examined as to the suitability of the habitat and space available for recovery. . . . If the existing habitat or space is not suitable, then the area will be eliminated from consideration. . . ."[40]

It makes little difference to federal agencies whether grizzlies occupy a living ecosystem or not. To recover, grizzlies must be fortunate enough first to live in "suitable habitat." Depending on who defines this term, there soon may be no such place left.

Decisions for the Future

In fall 1991, Jon Almack submitted his North Cascades grizzly habitat evaluation to an IGBC-appointed scientific committee. Almack's final report combined a map of North Cascades habitat classes (open-canopy forest, grass/forb meadows, etc.) with priorities for each class based on the seasonal availability of Craighead's seven habitat factors. Almack used a statistical Delphi system procedure to assign a numerical rating and scale to each habitat class.[41] By making aggregates of all the values for each habitat class he could determine if the North Cascades grizzly bear ecosystem provided poor, average, or excellent habitat for the animals.

Almack's data proved that the North Cascades has average to excellent grizzly habitat, and the IGBC declared the North Cascades to be a grizzly bear recovery area in December 1991. But with no further protection under the Endangered Species Act forthcoming and only a recovery progress report required by the end of 1992, doubt lingers as to the intentions of the committee and the federal agencies responsible for the bear. Contrary to the simplistic claims of Bruce McClellan, who suggests that proper logging practices combined with road closures are sufficient to protect grizzlies, the facts of bear biology guarantee nothing. As Almack knows, though the grizzly has been protected by the Endangered Species Act for sixteen years, little has been done to tailor management to the bear's needs in the Greater North Cascades.

The situation is as complex as the mythic relationship between humans and bears. Federal agencies must follow guidelines designed to ensure that land management does not place threatened or endangered species at greater risk. The Endangered Species Act and the National Environmental Policy Act require the agencies to complete biological assessments for major projects (dams, highways, mines, etc.) and biological evaluations for smaller-scale activities (most timber sales, trails, construction, and the like). In 1986, the IGBC drew up national grizzly bear-management guidelines that were approved by the secretaries of agriculture and the interior and became official policy. In order to implement these guidelines, most designated grizzly ecosystems have been divided into bear management units. But the Greater North Cascades, because it did not become an official recovery zone until recently, has never been divided. And there is no mention in the IGBC recovery-area decision of doing so. In addition, the Forest Service is working on a model that would describe the possible cumulative effects of proposed management activities on grizzlies. The model requires data that has not been collected in the North Cascades. This puts the smallest population of grizzly bears in the continental United States at a distinct disadvantage.

In 1988, Chris Servheen, recovery coordinator for the IGBC, asked federal agencies in the Northwest to use the national bear-management guidelines while Almack's research was being completed.[42] The Fish and Wildlife Service, the agency responsible for overseeing the Endangered Species Act, placed the grizzly on its list of regional species that must be evaluated in any land-management plan. Jon Almack also lobbied for interim bear-management guidelines for the North Cascades. In 1988, he drafted a set for agency biologists who write assessments and evaluations.[43] Almack entered all his bear-sighting data into the interagency computer system. But the IGBC management guidelines are still not being used in the Greater North Cascades, and interim rules have not been adopted. Few biologists use Almack's data, though it is just a finger tap away. My research at the office of the supervisor of the Mount Baker–Snoqualmie National Forest in Seattle has shown that, throughout the 1980s, the grizzly received only cursory treatment in the majority of biological evaluations written for timber sales in bear habitat.[44] Lack of data is often cited as a reason for logging to proceed—the onus is placed squarely on the animal, even as biologists cannot spring funding from the agencies to complete necessary research. In another typical pattern, the Wenatchee forest plan predicts negative impacts on the bear but outlines no protective strategy. And so the grizzly wanders, wraithlike and unshielded, through the high cirque basins and shrubfields of Bacon Creek, the hidden watersheds under Hozomeen, and the open forests of Teanaway Buttes.

Map	*Date*	*Observation*	*Reliability*	*Location*	*Reported by*
19	1989	1 Adult (photo)	3	Unknown	English
20	Apr 1989	Food Cache	1	Hozomeen	Almack
21	Apr 1989	Adults	2	Methow River	Nelson
22	May 1989	Tracks (photo)	2	Ross Lake NRA	Almack
23	31 May 39	1 Adult	2	Entiat	Heinle

This is how the grizzly appears in the flicker of an interagency computer screen—the main access a Forest Service biologist has to *Ursus arctos* in the North Cascades.[45]

The Endangered Species Act provides little legal clarification. Project-level consultations do not take place between the Forest Service and the Fish and Wildlife Service.[46] The Forest Service determines whether its own development projects would affect the bear—if not, the bear is dropped from consideration. The absence of an official regional recovery plan has given the agencies an excuse to do as little as possible until such a plan is constructed. We act as if we knew nothing about grizzlies from other ecosystems. We pretend that interim guidelines are not available. We act as if the grizzly bear did not exist.

In the shadow of Glacier Peak, evening slants down. There are no bear shadows in the basin. The trail toward camp lies deep in dust. A solitary Swainsons thrush sings evensong as I head back to camp. There I eat supper away from the tent and hang my food downwind, high in a mountain hemlock, for the night. I am not the same person I was when I first began exploring these steep mountains; the grizzly has changed for me as well. As in the Bear Mother myth, it still stands for renewal in the Greater North Cascades. But the bear continually invites us toward new meanings. As our biological knowledge of this animal has expanded to encompass ecosystem-scale relationships such as viable populations, cumulative effects, and biological corridors, the grizzly itself has come to define the sweep of a greater ecosystem beyond the bounds of the North Cascades.

Our management practice must now match the full range of the bear. As in the original myth, the issue of survival, for both the grizzly bear and ourselves, calls us to learn to "participate in coexistence."[47] Less logging, road closures, reduced grazing and hunting, and the immediate implementation of bear-management guidelines are the practical requirements for an appropriate relationship between bears and people in the Greater North Cascades.

We cannot save the grizzly bear. There is no guarantee that what we do for the bear now will be successful. But we can begin to act on its behalf. We, the grizzly bear, and the Greater North Cascades share a common fate; no one of us can be renewed without each of the others.

CHAPTER FOUR

LAWS ON THE LAND

Back/beneath the asphalt,
the parked tractors and stacked logs:

the ghosts of all the marmots
are gnawing at the roots of things.

Whole hillsides full of stumps
are getting restless.

Wolf spirits, Bear spirits
can't find work.

TIM MCNULTY,
Where All the Marmots from
Marmot Pass Went To

Sometimes, in certain special places, the animals return. In May of 1990 wolves were sighted near Hozomeen Lake in the North Cascades National Park complex, a few miles south of British Columbia. Upon investigation, government biologists located a den with mother and cubs. The wolf family responded to human howling and seemed settled in their new home. It was the first confirmed active den site for northern gray wolves discovered in the Greater North Cascades since 1930.[1]

Less than a month later, a second pack and den was found much farther south of the international boundary, near the upper end of the Methow Valley in the Okanogan National Forest, barely five miles from the site of the proposed Early Winters ski area. Like their kin to the north, this pack appeared established, active, and healthy. In August a third pack was discovered even farther south, near the crest of the Cascades in the Wenatchee National Forest. For all that we have done and left undone in the name of predator control and endangered species protection, these wolves were reclaiming, on their own, what was once incontestably theirs.

The northern gray wolf is listed as endangered under the Endangered Species Act. Since 1973, one of the world's most persecuted mammals has been protected by one of the most prohibitive of U.S. laws. The act, along with the Wilderness Act, the National Environmental Policy Act (NEPA), and the Clean Air and Clean Water acts, dates from the 1960s and 1970s, when Americans attempted to redress their abuse of the natural ecosystems of North America. For some 500 species gone extinct since the Pilgrims landed in 1620, these legal actions meant nothing. For the northern gray wolf, the grizzly bear, the northern spotted owl, and a host of other species, the jury is still out. As for ecosystem diversity in the Greater North Cascades, it is not at all clear that our legal efforts, so far, have made much of an impact at all.

When we lay a law on the land and native species, what does it demand of us? If the grizzly bear is listed under the Endangered Species Act and requires protection, why do we stumble to meet this task? When the northern gray wolf returns unannounced to its former range, why do debates ensue over whether it should be allowed to remain? It is unusual for us to ask these questions, let alone provide answers to them. Our response is bound to how we place ourselves in the world, on par with wolves and bears, or above them. One way Americans have addressed such questions is by passing laws to protect species, parks, and wilderness areas. But what if, when the laws *of* the land are laid *on* the land, the fit is so loose that elements of biodiversity fall through the legal safety net?

Is the Endangered Species Act working? Ask any grizzly bear or wolf in the Greater North Cascades. The laws we use to protect biodiversity, though they are touted as some of the most powerful in the world, are not doing the job. Since the act was passed in 1973, more species have gone extinct than have recovered to healthy population levels.[2] Since 1980, thirty-four species have gone extinct without the benefit of ever being listed.[3] Currently, out of a backlog of some 3,700 candidates, almost 1,000 species face extinction before they are listed.[4] Yet, as of 1991, funding levels permit the U.S. Fish and Wildlife Service to list only about fifty species per year—so few it would take the agency over seventy years to work through all current candidates. How is it that laws that supposedly protect biodiversity have come to epitomize Newmark's warning that the politics of protecting nature is incongruous with the ecological facts of life?

To solve this puzzle, we must first explore the shortcomings of existing legislation. Our search must not be confined, however, to the legal language of the U.S. Code and the Code of Federal Regulations. If we are fully to grasp our laws on the land we must not forget how rules and regulations are shaped by environmental values. Americans have always argued over what the good life is, an inquiry that has not often included species other than ourselves. As we examine our laws for "ecological congruency," we need to keep in mind two questions: Why did we shape our laws in this fashion, and how might future laws be shaped to protect *all* elements of biodiversity?

The Limits of Legal Preservation

There are three strands of law to explore in unraveling our legal response to biodiversity issues. Overarching all the statutes specific to wildlands and wild creatures is the National Environmental Policy Act (NEPA), passed in 1969. Though there was little debate upon its passage, NEPA is one of the most revolutionary laws ever approved by Congress. It opened up the entire range of federal land planning to public participation and judicial review by instituting the now familiar Environmental Impact Statement (EIS) process. NEPA democratized federal land planning and as a result (unforeseen) changed the culture of federal land-management bureaucracy. To produce an adequate EIS, a federal agency proposing a project must first document the potential effects of a range of alternatives. Then it must compare and contrast the environmental consequences of each alternative and predict how each decision fits into the larger context of the cumulative effects of other projects in the vicinity of the original proposal. Without hiring new wildlife biologists, hydrologists, landscape architects, and others, no agency could hope to meet NEPA's data demands. This influx of new blood diversified opinion within the agencies, a trend that was reinforced through the interagency-consultation and public-involvement mandates of the law.

Yet NEPA, for all its significance, is primarily a procedural law. It does not mandate environmental protection per se, it only ensures environmental review before major land-use decisions are made. To investigate the full spectrum of legal approaches to biodiversity we must move from procedural requirements toward laws passed to protect wilderness, wildlife, and disappearing species.

Most American wildlife laws are designed to protect species or groups of species. Laws protecting places (national parks, refuges, and so forth) have been passed, though the rationale behind them, more often than not, is sustaining "human use and enjoyment," not wildlife per se. These two kinds of law reveal fundamental flaws in our cultural image of nature as well as problems with the statutes themselves. We discount the connection between species and habitat even as we disregard the link between ecosystems and ourselves. According to our (legal) world view, animals do not depend on habitat and we do not depend on nature.

This dualism is found throughout the national parks movement, which can be considered the United States's initial legal foray into biodiversity protection. But, in fact, the 1872 Yellowstone Park Act had little to do with protecting wild animals or wildlands. Instead, the Yellowstone region was withdrawn to "provide for the preservation . . . of all timber, mineral deposits, natural curiosities, or wonders within the park."[5] Historian Alfred Runte has put it plainly: "The national park idea evolved out of a concern for natural wonders as monuments rather than from an appreciation of the value of landscape in its broadest sense, both animate and inanimate."[6] The purpose of national parks was revealed in the 1916 Park Service Organic Act as a two-pronged commitment to "conserve the scenery and the natural and historical objects and the wildlife therein and to provide for the enjoyment of the same. . . . "[7] No national-park legislation placed biodiversity or wildlands first until the 1934 law that created Everglades National Park: "The said area or areas shall be permanently reserved as a wilderness . . . (and no development) must interfere with the preservation intact of the unique flora and fauna and the essential primitive conditions."[8] Unfortunately, up through today, this has been the exception rather than the rule.

Our growing understanding of how nature works has not yet been translated into park legislation. In 1978, Organic Act amendments that focused on Redwood National Park rescued the northern California redwoods in the original, poorly designed park from erosion caused by clearcutting outside the park boundary; this may augur future change. According to some scholars, this law contains language that gives the Park Service authority over land use adjacent to *any* national park if such use would "derogate" park values and purposes.[9] But the Park Service, not wishing to step on the toes of its neighboring agencies, has never attempted to assert this potentially powerful mandate.

Politics, not ecology, still controls boundary marking. To avoid political pitfalls, the 1968 law that established the North Cascades National Park complex split the area into four different administrative units, only one of which is actually a national park. Twenty years later, when a bill was being drafted to classify new wilderness areas in the North Cascades Park complex and Mount Rainier and Olympic National Parks, a section called for an ecological study of the three reserves. The five-year study would have been a conservation biologist's dream, addressing "the ability . . . of each park to preserve and maintain fully functioning ecosystems" and to "maintain healthy populations of all species native to the parks," evaluating "impacts to parks from uses on surrounding lands," and determining "where the parks are not large enough to provide or whose shape prevents full protection for park resources and values. . . . "[10] The draft even called for an evaluation of the most representative, unaltered watershed associated with each park that flowed from timberline to tidewater, "not just a narrow riverine corridor," so that an entire valley might be considered for protection. Such "radical" concepts, however, were not to survive congressional compromise. To ensure passage of the wilderness classifications, the ecological study was dropped from the final draft of the law.

The 1964 Wilderness Act, a law commonly thought to protect biodiversity, is built on a similar foundation of preservation for "use and enjoyment." The act has the most lengthy and convoluted history of any U.S. law. Through nine years and sixty-six rewrites, the strong protective measures of the original draft were compromised severely. The act's definition of wilderness allows that it "may also contain ecological . . . features . . . of value," but in the context of the law's recreation-based implementation, this is clearly of secondary interest.[11]

Both parks and wildernesses do protect elements of biodiversity. But protected areas were not designed by Congress for this purpose, and we must not expect them to perform an adequate job. Conservation biologist Reed Noss has discovered that the National Wilderness Preservation System does a poor job indeed. Concerned about Newmark's warning, Noss evaluated how many U.S. ecosystem types, following the Bailey-Kuchler classification system, were protected in wilderness areas of over 250,000 acres. He found that only 19 percent of 261 ecosystem types in the United States and Puerto Rico were represented.[12] Other researchers have found that the national park system fails to represent 33 percent of U.S. ecosystem types of *any* size.[13] In the North Cascades, Jon Almack's ecosystem core contains almost all of the region's ecosystem types. Parks and wilderness areas comprise almost 35 percent of the entire region. These figures attest to the health of the region relative to others around the country. But grizzly and gray wolf populations are not at all healthy in the Greater North Cascades. How can biodiversity flourish in one place and be at risk in another? To answer this question, we need to explore American statutory protection for wildlife.

The Endangered Species Act and Biodiversity

Wildlife law in the United States is rooted in Old World rule, when kings exerted control over animals. Regal authority was carried to the

New World by our ancestors and invested in the states and territories.[14] Federal jurisdiction over wildlife did not become established until the passage of the Lacey Act in 1900. This law, which gave the secretary of agriculture the power to enforce state wildlife laws, was passed in response to the extinction of the passenger pigeon, a prominent game bird in the nineteenth century. From this beginning the federal government has broadened its statutory authority based on its power to control interstate commerce and to make and enforce international treaties and the simple fact that 30 percent of the United States is federally owned.[15]

The first national wildlife refuge was created on Pelican Island, Florida, in 1903, followed soon thereafter by the designation of the National Bison Range in Montana in 1908. These reserves were established to protect two American species that had undergone dramatic population declines. But it was not until the passage of the 1929 Migratory Bird Conservation Act that the government moved to create a system of "inviolate sanctuaries" for wildlife. (Amendments and laws passed since allow hunting and other consumptive activity in these sanctuaries.)

By the early 1960s, industrial development in the United States was outstripping the growth of parks and reserves even as it outpaced the implementation of the wildlife laws that were on the books. At the same time, scientific understanding of extinction and public awareness of environmental degradation were on the rise. The result was a series of congressional actions meant to strengthen wildlife protection.

In 1964, the same year the Wilderness Act passed, a committee of scientists in the Interior Department produced the Redbook, the first federal list of vertebrate species thought to be endangered. Congress soon followed suit with the passage of the 1966 Endangered Species Preservation Act. This prototype of more powerful laws to come protected only vertebrates, was to be implemented only to the extent "practicable and consistent" with established policies, and did not prohibit the killing of listed species except in limited geographical areas. In 1969 Congress replaced the 1966 legislation with the Endangered Species Conservation Act. This new law extended protection to invertebrates but did little else for U.S. species. However, it did require the secretary of the interior to draw up a list of international species threatened with extinction, and it called for a global convention on protection and trade in species thought to be at risk.

The new law did not prove sufficient in the face of escalating environmental degradation. In 1972 Congress enacted the Marine Mammal Protection Act, which, for the first time, placed penalties on "taking" (harming) listed species and also protected depleted as well as endangered populations. These were important steps toward a more comprehensive approach, but the law only covered ocean-dwelling mammals. In early 1973, when the long-delayed Convention on International Trade in Endangered Species resulted in worldwide trade restrictions, it became Congress's task to bring domestic policy into line with the new international regulations, placate the growing environment movement, and, finally, create a sweeping law that would fully protect endangered species.

That law became the Endangered Species Act of 1973. Its purpose, even after three major amendments (1978, 1982, 1988), is to identify species in danger of extinction, to protect those species from "jeopardy," and to foster their recovery to adequate population levels. The law is strongly prohibitive: "If species are biologically endangered, they must be listed. If they are listed (except for plants), no one can harm them and agencies must act to protect them. If any individual/organization misbehaves, they/it can be sued. At legislative face value, there is no room for negotiation or discussion. . . ."[16] The act defines the words *endangered* and *threatened*, details the listing process, makes provisions for habitat protection and acquisition, restricts the taking of listed species, mandates species recovery, links domestic policy with the Convention on International Trade in Endangered Species, outlines interagency cooperation, and carries enforcement penalties. Once a species is listed, the goal of the act's protection and management regulations is to get the species off the list as quickly as possible.

However, as the case of the grizzly bear and gray wolf in the Greater North Cascades might well come to illustrate, extinction could be the most common delisting device. The Endangered Species Act is riddled with loopholes large enough to allow the world's biggest omnivore and most wide-ranging predator, as well as thousands of other species, to teeter on the brink of forever. Listed species are up against the edge because of the following:

Inadequate scientific understanding of extinction. The act was written almost a decade prior to the rise of conservation biology as a discipline. Though the law requires the "best biological information" to be used in making decisions, this is problematic when the structure of the law itself does not reflect current scientific knowledge.

Lack of congressional clarity. While the overall intent of the act is clearly prohibitive, Congress has shown its ambivalence toward species protection in key sections of the original law as well as in recent amendments.

Conflicting regulatory interpretations. Any new U.S. law must first be interpreted by the administration and codified in the Code of Federal Regulations. The secretary of the interior does this for the Endangered Species Act. If the administration is of a different political persuasion, the regulations may not follow congressional intent.

Poor agency implementation. In turn, specific federal agencies such as the Forest Service must implement the Code of Federal Regulations. This can also weaken the letter of the law because the agencies are the action arm of an Administration, even as they are beholden to their own unique agendas.

Because of these four problems, the Endangered Species Act is not capable of solving species-level biodiversity dilemmas, let alone habitat- or ecosystem-scale problems.[17] Here, to provide the reader with a guide, are ten of the law's most glaring defects (section numbers refer to the Endangered Species Act):

1. The act plays taxonomic favorites, giving animals stronger protection than plants (section 9). This has no justification in conservation biology: In ecosystems, plants may be just as important as any other species. Favoritism has been translated by the agencies into funding decisions that benefit only a few dozen high-profile species. To address this inequality (but without dealing with the underlying problem), Congress, in the 1988 amendments, required that agency efforts proceed "without regard to taxonomic classification."[18] But the problem remains. In 1990, the ten species receiving the most monies accounted for 49 percent of all endangered species funding.[19] All ten were vertebrates.

2. The law allows subpopulations of listed species to be treated separately (section 4). The grizzly bear in Greater Yellowstone can be listed as endangered even if those in the North Cascades are (merely) threatened. But Congress left open to interpretation exactly what sort of subpopulation could be defined as a population segment, that is, do all continental U.S. grizzlies count as one segment, or can each of the six distinct grizzly ecosystem populations be considered? This is an extremely important question because evolutionary biology suggests the importance of protecting genetically distinct subpopulations below the species level.[20] But because too many listed subpopulations would reduce administrative "flexibility," the secretary of the interior has ruled that he can define subpopulations at his discretion. So far, he has only allowed ten species to be considered: the bald eagle, peregrine falcon, grizzly bear, red-cockaded woodpecker, brown pelican, sea turtles, ocelot and jaguarundi, and piping plover. Given the effects of habitat fragmentation, this short list is biologically indefensible.

3. One of the fundamental goals of the Endangered Species Act is "to provide a means whereby the ecosystems upon which endangered species and threatened species depend, may be conserved."[21] This is perhaps the most controversial provision of the statute. In 1978 Congress amended the law to require a critical-habitat designation for each species at the time of listing. This threw the listing process into a state of shock from which it has yet to recover. Two thousand species proposed for listing were immediately withdrawn, few species were listed in the next several years, and today there is a continuing struggle to get out from under a massive backlog of candidate species.

Why did this happen? Common sense as well as conservation biology suggests that a viable population needs viable habitat. The act defines critical habitat as

> (i) the specific areas within the geographic area occupied by the species, at the time it is listed . . . on which are found those physical or biological features (I) essential to the conservation of the species and (II) which may require management considerations or protection; and (ii) specific areas outside the geographical area occupied by the species at the time it is listed . . . upon a determination by the Secretary that such areas are essential for the conservation of the species.[22]

Prior to the 1978 amendments, land managers had the luxury of working to protect species without having to worry about protecting their habitat directly. Species could be divorced from their environment. Critical-habitat designation changed all that by tying species protection to specific places.

But Congress, concerned that this would be seen as too restrictive, weakened the amendment by allowing the secretary of the interior the discretion not to designate critical habitat for species listed before 1978. Lawmakers also allowed the secretary to declare current and future critical habitat only "to the extent prudent and determinable." As if this was not enough to weaken the new rule, Congress allowed the secretary to use economic impacts as reason to reduce or eliminate such designations. As Daniel Rohlf has summarized, this "is a statutory aberration at odds with the purpose and structure of the remainder" of the Endangered Species Act.[23]

As might be expected, few critical habitats have been so designated. In 1986, forty-one of forty-six listed species were not tied to their critical habitat.[24] Though Congress has amended the act twice to speed up the listing process, they have never addressed this fundamental flaw in their handiwork.

4. Section 7 contains powerful provisions to protect listed species from jeopardy. It also outlines how all federal agencies are to consult with the Fish and Wildlife Service and the National Marine Fisheries Service (NMFS), which administer the Endangered Species Act. Still, this section of the law does not serve species as well as it could.

The trouble begins with Congress. The legislation is unclear as to whether the word *jeopardy* refers to actions that would simply limit a species recovery or threaten its very survival. From the standpoint of conservation biology the issue is not the continued survival of a population, which could well be accomplished in a zoo, but population viability. The act uses a narrow legal definition of viability ("continued existence"), not a biological one. This has permitted the secretary of the interior to write regulations that interpret jeopardy to mean survival, not recovery per se. Because of this, the Fish and Wildlife Service is not bound to label the North Cascades subpopulation of grizzly bears as an official recovery ecosystem even though this subpopulation is likely the most endangered of them all. This loophole undermines the entire *raison d'être* of the Endangered Species Act.

Critical-habitat designation is also linked to the survival/recovery issue. Under section 7, federal activities cannot destroy or modify the critical habitat of a listed species. But the secretary's regulations follow the same jeopardy interpretation: If the survival of the species is not directly threatened by loss of critical habitat, the agency may proceed even if recovery is compromised. This interpretation lies behind the Fish and Wildlife Service's approval of a plan for California's Coachella Valley fringe-toed lizard that will eliminate 75 percent of the reptile's habitat in favor of housing development. Of course, if no critical habitat is designated, then an agency doesn't have to worry about the fine points of whether an action might involve survival or recovery.

5. Consultation between a federal agency proposing to take action and the secretary of the interior (through the Fish and Wildlife Service and the NMFS) is the procedural heart of section 7. This is where agencies must integrate their daily activities with the Endangered Species Act. But the law and the secretary's regulations conflict in major areas of the process. The act requires the Fish and Wildlife Service and the NMFS to prepare regional lists of threatened and endangered species. If a listed species "may occur" in a proposed project area, the action agency must prepare a biological assessment to determine how the species could be affected. The project cannot proceed while the assessment is being prepared and evaluated. If the biological assessment suggests a negative impact, then the agency must begin formal consultation with the Fish and Wildlife Service or the NMFS. This results in approval, mitigation, or denial of the proposal, following what is best for the species in question.

The regulations interpret the law differently. The secretary has tied preparation of biological assessments to whether a project is "major" or "minor" following definitions in the NEPA.[25] (Under the NEPA, for instance, most timber sales are minor actions.) Because these definitions are unclear, the agency involved has discretion in this important decision. The regulations also allow the action agency to determine, following completion of a biological assessment, whether the proposal is likely to affect a listed species or its habitat, and the ESA does not specify what the contents of an assessment should include. This is tantamount to asking the fox to watch the chickens, since it is the agency that wants to implement the proposal. No court decisions have clarified these procedural ambiguities.

6. The distinction between formal and informal consultation is not academic. When agencies formally consult with the Fish and Wildlife Service or NMFS under section 7, they are required to use the "best scientific and commercial data available." But they do not have to use such data explicitly during informal consultations. This loophole allows an agency to use less than the best scientific information in preparing biological assessments.

7. Even the use of the best scientific data available does not guarantee protection for threatened and endangered species. Lack of data and uncertainty cast a pall over the lives of many, if not most, listed species. In the North Cascades, the best data we have on grizzly populations is that there may be ten to twenty animals living somewhere in an 8-million-acre haystack. When Congress wrote the Endangered Species Act it expected that solid data would be available with which to make decisions. Congress also assumed that science would yield incontrovertible results. But data is almost always ambiguous. And the act provides no guidance as to whether a project can proceed while further data is being collected. The secretary's regulations, on the other hand, allow the action agency to decide whether data is sufficient or not.[26]

8. Development activities do not take place in a vacuum. Section 7 requires the agencies to study the cumulative effects of proposals. Evaluating each timber sale, one at a time, in a watershed that was to be entirely cut over twenty years would do little to protect a threatened orchid. Again, the secretary's regulations fall short of the spirit of the act. The regulations do not bind together past, present, and future actions. In 1981 they were loosened to include only *currently* proposed federal actions in cumulative-effects analyses. This means that if the Forest Service plans to build only thirty miles of new road in high-quality grizzly habitat over the next ten years, only those four miles of road associated with a current proposed sale need be analyzed. As they have been interpreted, the regulations make a mockery of the concept of scientific cumulative effects.

9. In the 1988 amendments to the Endangered Species Act, Congress required the Fish and Wildlife Service and NMFS to draw up plans to help with listed species recovery. These recovery plans, with some exceptions, have been about as successful as critical-habitat designation. As of 1989 less than 55 percent of listed species had recovery plans.[27] Only 3 percent of listed species were actually recovering.[28] Policy analysts Robert Culbert and Robert Blair at the University of Michigan have found political motivations in the writing and implementation of the plans.[29] The biology in several of the plans is suspect. For example, the recovery plan for the northern rocky mountain wolf flies in the face of viable-population theory by defining a "recovered" population as ten breeding pairs over three years' time. In the Greater North Cascades, the wolf has no recovery plan and the Fish and Wildlife Service, as of this writing, has not initiated a study to design one. And since the 1980s, when the Reagan administration chose to block funding for recovery plans, it has been a difficult political battle to implement the amendment at all.[30]

10. While the Endangered Species Act focuses on single species, it is primarily a federal-lands-specific statute. Both goals prevent the act from protecting biodiversity at the landscape scale. Roughly 80 percent of the United States, outside of Alaska, is in state and private hands. Section 7 jeopardy standards and section 4 critical-habitat designation do not apply to these lands. As Betsy Parry of the University of Wisconsin has observed, "A private owner can erase a virgin forest or prairie from his land without even filling out a form, even when endangered species are present. . . ."[31] Even the "no takings" provision of the act (section 9), which does apply to nonfederal land, is subject to interpretation. No takings (that is, harm) may not refer to delaying the recovery of a listed species. The recent conversion of 10,000 acres of native vegetation into vineyards by Gallo, Inc., in Santa Barbara County, California, in what could have once again become California condor habitat, underscores this point.

Scrutiny of the Endangered Species Act brings an overarching problem to light. Because the law does not link the way we use nature with protecting biodiversity, species are predestined to become threatened and endangered. The act is not so much for species as against extinction.

Under the law, we can only react to the inevitable as human population and resource consumption increase. In the face of so-called progress, and with the act's provisions triggered only by the conditions of extinction, our efforts are doomed to fail.

Given the general unwillingness to address the real causes of the biodiversity crisis, the specific shortcomings of the Endangered Species Act may appear superfluous. But the details of the law provide valuable insight. We can now comprehend why gray wolves and grizzlies, in one of the largest wildlands in the continental United States, are barely hanging on even after almost two decades of protection. We are able to puzzle out why the law affords little protection for species habitat. This, in turn, illuminates the cultural confusion behind species-specific laws like the Endangered Species Act, which focus on species and not their habitat, and "sublime-place" legislation like the National Park and Wilderness acts, which have created reserves that do not protect the species that call them home. The details also spotlight how our scientific understanding of biodiversity, so far, has played but a minor role in legislation.

The Endangered Species Act is a good example of how biodiversity theory mixes with politics. Congressional attempts to amend the law have been ambivalent. The secretary of the interior's regulations, extremely narrow and oftentimes counterproductive, are the product of politics leavened but little by biology. We are left pondering the rift between human values and the natural world, and the relationship between the letter of the law and the protection of place.

The National Forest Management Act

Laws, like human images of nature, are subject to change. In 1976 Congress attempted to protect biodiversity in full, at least in the national forests.

The National Forest Management Act (NFMA) is the only U.S. law that specifically requires a federal agency (the Forest Service) to "provide for diversity of plant and animal communities. . . ."[32] The NFMA is supposed to provide both species- and ecosystem-level protection on

Forest Service lands. But is the law adequate? How much of it is grounded in conservation biology? Has it been interpreted properly? Is it being implemented? How has the law been viewed by the courts? Understanding the NFMA is one key to our search for an adequate legal standard by which to protect biodiversity.

The NFMA is a complex law.[33] It addresses most facets of national-forest management through comprehensive planning. Biodiversity is but one concern among many. To understand the NFMA, and specifically its biodiversity provisions, one must first explore the reasons Congress passed a law that resulted in "the strongest environmental and silvicultural controls ever imposed by any legislation dealing with the national forests."[34]

Up until 1976 two laws, the 1897 Organic Act and the 1960 Multiple-Use/Sustained-Yield Act (MUSY), guided the Forest Service. The 1897 law, a relic of the nineteenth century, directed the Forest Service to manage for water quality and a continuous supply of timber. The agency was given the discretion to achieve these general goals as it best saw fit. Such discretionary authority was confirmed by the passage of the MUSY. The new law was formulated in response to increasing pressures on national forests in the decade following the end of World War II. Between 1950 and 1959, annual timber cut off national forest-land more than doubled while recreational visits tripled.[35] America was booming. But industrial expansion inevitably led to conflict—people don't want to camp in clearcuts. With the MUSY, Congress pushed the Forest Service to balance management in the interest not only of wood and clean water but of wildlife, fish, recreation, and grazing as well.

The congressional nudge was gentle. The law, drafted by the Forest Service, did not demand strict equity between different management concerns. It only required equal "consideration" of multiple uses, leaving the determination of the balance to the discretion of the agency.

Since passage of the MUSY, the Forest Service, contrary to the spirit if not the letter of the law, increased its commitment to timber production in national forests. The decade of the 1960s saw cutting rise by another 30 percent. The agency also increased its planning for other forest activities by experimenting with land-use zoning and wildlife, recreation, and other use-specific management plans. These efforts were guided by administrative rules gleaned from the Forest Service manual, but they were not legally binding. The few court challenges to the Forest Service's preference for logging failed—the MUSY was simply not prescriptive enough. Even the passage of the Wilderness Act, the Wild and Scenic Rivers Act, and the NEPA did not divert the agency from its primary focus.

But the new environmental legislation augered changes in societal values that even the Forest Service could not resist forever. And what citizens demanded, Congress might eventually provide. With clear-cutting at an all-time high and the environmental movement in full swing, the only thing lacking was a match for the tinder.

In 1975, a West Virginia court injunction provided the spark. In response to a citizen's group law suit, the Forest Service was enjoined by the U.S. District Court from using clear-cutting to log. Since this was the predominant technique in use, the Monongahela Decision brought timber cutting to a screeching halt throughout the entire national forest system and sent the Forest Service to Congress for remedial legislation. This time, however, the agency was in no position to write its own legislative ticket. Congress had to acknowledge the prolonged public outcry against clear-cutting and the agency's timber-heavy agenda that had been building over the years.

The new law, the NFMA, was passed in October 1976. It may yet force substantive change on the Forest Service. That it has not yet done so, even after fifteen years, illustrates the hazards of congressional compromise, the impediments to change within an entrenched Forest Service bureaucracy, the slow, incremental growth of scientific knowledge, and the continuing evolution of ideas about nature as humans grope toward a reckoning with environmental constraints.

There is little doubt that Congress intended the NFMA to reform Forest Service land management in a comprehensive way. Though the law specifically sanctions clear-cutting, it also sets limits on a broad array of logging practices, requires inventories for all forest resources, and mandates public participation in a systemwide, interdisciplinary planning process. Each national forest must produce a detailed forest plan for a ten- to fifty-year period. Once completed, the plans become legally binding, and citizens have the right to sue the Forest Service if those plans are inadequate. This is the most significant provision of the NFMA: It opens up Forest Service decisions to judicial review.

Congress, nevertheless, was not about to strip the Forest Service of all professional discretion. One of the main weaknesses of the NFMA is that, like the Endangered Species Act, it is frequently couched in vague, discretionary language open to broad interpretation. This belies the rhetoric that surrounded the law's passage. Senator Hubert Humphrey, whose bill became the building block for the NFMA, proclaimed: "The days have ended when the forest may be viewed only as trees and trees viewed only as timber. The soil and the water, the grasses and the shrubs, the fish and the wildlife, and the beauty that is the forest must become integral parts of resource managers' thinking and actions."[36]

Yet Congress hedged its bets in important sections of the act. The NFMA, as a result, is a law awaiting judicial interpretation on so many fronts that it will take years to clarify what lawmakers could have made more plain from the beginning.

The diversity section of the NFMA is a prime example of Congress's lack of resolve. This is the slender thread upon which hangs the protection of the full spectrum of life in national forest ecosystems. Section 6(g) 3(B) requires the Forest Service to "provide for diversity of plant and animal communities based on the suitability and capability of the specific land area in order to meet overall multiple-use objectives, and within the multiple-use objectives of a land management plan adopted pursuant to this section, provide, where appropriate, to the degree practicable, for steps to be taken to preserve the diversity of tree species similar to that existing in the region controlled by the plan."[37]

This ambiguous passage does not define diversity or suggest to what extent it is to be protected. The second half of the passage gives biologically uncalled-for weight to tree species, while the first part is qualified by the phrase "in order to meet multiple-use objectives." Though the legislative history of the NFMA clearly shows that Congress was concerned with the overall ecological health of forests, it also reveals that lawmakers were at least as interested in preventing the spread of forestry monocultures as in protecting biodiversity per se.[38] Either Congress did not understand biodiversity well enough to act decisively, did not care to do so, or both.

Fortunately, through the NFMA, Congress also required the secretary of agriculture to convene a committee of scientists to write implementing guidelines, "modeled" on the law, that would provide more specific direction to the Forest Service. Once adopted, these NFMA regulations would become part of the Code of Federal Regulations and have the force of law. It was up to the committee of scientists, which completed its work in 1979 and approved subsequent revisions in 1982, to put scientific meat on the NFMA skeleton.

The committee did just that. Title 36, section 219 of the code represents the strongest manifesto in favor of biodiversity that exists in U.S. law. First, the committee wanted NFMA regulations to "go beyond a narrow and limited restatement of the language of the [act] to assure that the Forest Service shall indeed 'provide for diversity' by managing and preserving existing variety."[39] To this end, the regulations require the agency to "preserve and enhance the diversity of plant and animal communities so that it is at least as great as that which would be expected in a natural forest," to recognize national forests as ecosystems and consider ecological relationships, and to assess the effects of proposed management practices on diversity throughout the planning process.[40]

The regulations are not without problems, however. They define diversity as "the distribution and abundance of different plant and animal communities and species within the area covered by a Land and Resource Management Plan."[41] By accenting only species and communities, this definition excludes the genetic and landscape aspects of biodiversity. Diversity is affected by individual forest plans, which cover administrative, not ecological, boundaries. And the word *enhance* can be interpreted to give the Forest Service the discretion to "improve" on natural ecosystems to suit production goals.

Though the committee of scientists neglected important levels of biodiversity, it fine-tuned species-level regulations with four substantive requirements governing viable populations, indicator species, scientific inventories, and species monitoring. The regulations require that fish and wildlife habitat be managed to maintain viable populations of existing native and desired nonnative species. A viable population is defined as "one which has the estimated number and distribution of reproductive individuals to insure its continued existence and is well distributed in the planning area." Unlike provisions in the Endangered Species Act, "habitat must be provided to support, at least, a minimum number of reproductive individuals. . . . "[42] The committee made it abundantly clear that, at least throughout a single national forest, each native species must be protected, sufficient habitat must be tied to protection, and the spatial relationship (distribution) of populations must be accounted for.

But hundreds of vertebrates get their living in each national forest. It would take decades of field research to gather enough data on all of them to implement the committee's viable-population rules. Recognizing this problem, the committee leaned heavily on the management indicator species (MIS) concept. This theory suggests that certain species indicate the general health of ecosystems in such a way that a decline in their population signals decline for the rest of the species with which they share habitat. The concept assumes an overlap in different species' living requirements, an assumption that not all ecologists agree with.[43]

If indicator-species selections are based on scientifically credible analyses, then, theoretically, they can play their canary-in-the-coal-mine role. But how would the Forest Service choose representative indicators from the hundreds of species present in each forest?

The committee of scientists narrowed Forest Service selections by giving the agency five categories of species to choose from "where appropriate": threatened and endangered species, sensitive species, game and commercial species, nongame species of interest, and ecological indicators. Though the phrase "where appropriate" allows the Forest Service some discretion, other language in this section restricts the indicator-species concept. After selection, planners must "establish objectives for the maintenance and improvement of habitat," and these must be tied to both qualitative and quantitative habitat values as well as population trends for each indicator.[44]

There was little room for the Forest Service to misapply the MIS concept until 1982. During revisions of the NFMA regulations, the committee of scientists added the phrase "to the degree consistent with overall multiple-use objectives" to the relevant MIS section. The Forest

Service was concerned that no population changes whatsoever would be allowed to result from proposed management actions under the original regulations.[45] The agency's successful lobbying campaign points to the significance of the indicator-species concept. In their definitive review of the NFMA, Charles Wilkinson and Mike Anderson spotlight this key section of the law: "The success or failure of wildlife and fish resources planning will depend largely upon the manner in which individual national forests choose their MIS." And so, like many other parts of the NFMA's biodiversity provisions, the MIS section cannot be clarified without judicial review of the agency's decisions.

Any analysis of viable populations and indicator species is only as good as the scientific data upon which it is based. So the committee of scientists required the Forest Service to conduct inventories of wildlife that "include quantitative data making possible the evaluation of biological diversity in terms of its prior and present condition."[46] It is clear from the regulations that a simple description of wildlife would not satisfy the law. If biodiversity was to gain equal footing with logging, then the best available data "of a kind, characteristic, and quality, and to the detail appropriate for management decisions" would have to be gathered.[47]

And the committee asked even more of the Forest Service. Wildlife population trends are, by definition, moving pictures, because data must be gathered over time as animals are born, grow old, die, and are affected by ecosystem change. A chosen indicator must be monitored to determine its response to logging, road building, and the other management activities. The NFMA regulations require that the Forest Service track "population trends . . . and relationships to habitat changes" for all MIS.[48]

These inventory and monitoring regulations place a heavy burden on the agency. The Forest Service has never been known for the quality of its baseline-resource inventories. And the committee of scientists was asking for research that in many cases had not been done. In 1976 the northern spotted owl, let alone the host of lesser-known species like the marbled murrelet and pine marten, was not a household name. The committee anticipated problems with the regulations: "Even if a federal government-wide crash program of data acquisition and storage were to begin tomorrow . . . the Forest Service cannot remedy decades of national indifference towards basic research data. . . ."[49] Nevertheless, the committee did not back away from requiring a solid scientific basis for biodiversity protection.

The NFMA regulations comprised the initial wave of the biodiversity revolution that broke over the Forest Service, but they missed deeper currents of change. The NFMA was Congress's tepid response to citizen outrage over Forest Service mismanagement. Public concern, in turn, was born of the broad environmental movement of the 1960s and 1970s, which gave voice to a common need for healthy ecosystems. The NFMA's passage was but another step in the shift of American attitudes in the direction of favoring ecosystems over resources. What is revolutionary about the act is the unusual clarity with which the committee established a scientific basis for biodiversity protection, the opening up of forest planning to the public, and the judicial oversight that bolsters these two sections of the law.

Of course, the Forest Service's world view collided with the new wave of American environmental consciousness, the NFMA, and the committee of scientists' regulations. With its tradition of timber first, the agency could not be expected to embrace these changes, even if it had been primed with the data and personnel to do so. As it stands, the NFMA regulations are pushing the Forest Service into the world of conservation biology and ecological theory. The agency has been forced to gear up its professional staff to deal credibly with viable populations, scientific inventories, and monitoring. And, because of the threat of judicial review, the Forest Service can no longer hide poor science behind the old smokescreen of managerial discretion.

Profound change occurs over time, and the revolutionary effects of the biodiversity crisis may be no exception to this rule. Forest plans have only been implemented in the last several years. Because legal challenges cannot be filed without a final plan to argue against, the NFMA has just begun to be interpreted by the courts. One does not need a court decision, however, to measure the profound threat to biodiversity that the first round of forest plans represents. Given the clarity of the NFMA regulations, it is disturbing to detail how the Forest Service has misinterpreted the protection of biodiversity. This problem is serious and widespread, one measure of the agency's reluctance to enter the modern world of ecological land management. In their study, Wilkinson and Anderson expressed a sanguine belief in the "essential wisdom" of the NFMA planning process. They saw the act as creating valuable inventories, engaging the public, and "holding out the promise of creating ordered and principled decisionmaking." A glance at the forest plans it shaped proves otherwise.

National Forest Planning

Plastic stickpins, colored felt-tip markers, Forest Service planimetric road maps, hours of tracing roadless area boundaries and sorting through timber-sale proposals—I had not expected to become so familiar with these tools and tasks when I moved to Bellingham, Washington, and the Greater North Cascades ecosystem in the summer of 1987. But they soon became essential to my primary occupation: curbing the destruction of old-growth forests.

As a college-age environmentalist in 1976 I had dutifully appealed to Congress to put prescriptive teeth into the NFMA. When lawmakers failed to do so, I wrote off the law with that mixture of frustration and resignation peculiar to youth. I expected the Forest Service to comply with the NFMA's underlying message. I was naive enough to count on both the agency and the future.

The Forest Service's brand of interdisciplinary management struck home in 1987 when I read *Forests for the Future?: A Report on National Forest Planning*. Compiled by the staff of the Wilderness Society, it was an indictment of the agency's nationwide planning goals.[50] It collected data from 107 of the 123 forest plans and gave a vivid account of exactly how the agency hoped to implement the NFMA over the next fifty years. The picture was not pretty. The Forest Service proposed an increase in logging of 72 percent. This would require the construction of 200,000 miles of logging roads, a system almost four times longer than the interstate highway network. Much of this logging and road building was planned in the de facto biodiversity buffer zones of the remaining unprotected, roadless areas of the forests. Only 17 percent of Forest Service roadless lands were recommended for wilderness designation.

My concern increased after I arrived in the Greater North Cascades. Here I found forest plans detailing activities that would result in the destruction of places and species I knew and loved: northern spotted owls, lowland Douglas-fir forests, the Cascade River, the Twentymile-Thirtymile area of the Okanogan highlands, and many more. Here in this part of Region 6 (Washington, Oregon, and northern California), the most timber-rich of Forest Service areas, only 45 percent of the unprotected wildlands in the North Cascades would escape development fifteen years into the future.[51]

The Greater North Cascades was under attack and needed defenders. Three tasks demanded immediate attention. First, plans for the three North Cascades national forests as well as the national park complex had to be reviewed. What did each plan propose to do with a forest's storehouse of biodiversity? How did national park lands fit in? Most important, what landscape-level picture would emerge when all the plans were evaluated together? Such a review could not be conducted without a solid grounding in conservation biology. My second task was to immerse myself in the scientific literature on landscape ecology, habitat fragmentation, island biogeography, patch dynamics, wildlife corridors, new forestry, and more.

To perform these first two tasks, I needed to become familiar with the modus operandi of the Forest Service and meet the staff biologists. The chain of decision making linking the field, the supervisor's office, the regional office, and Washington, D.C., had to be made clear. Timber sales most threatening to biodiversity would have to be identified and strategic plans laid to prevent them from occurring.

This last question is what led me to stickpins and colored pens. I traced the roadless remnants of diversity onto maps and jabbed markers into each section where timber sales were to be offered. My contributions were but a small part of a larger effort unofficially coordinated by Mitch Friedman of the Greater Ecosystem Alliance. The alliance had recently been created to "encourage a new level of ecological understanding of Pacific Northwest issues through a combination of conservation biology and activism." This quintessential low-budget outfit depended upon the energy and visionary savvy of Friedman and other concerned citizens who had read their local forest plans and seen what they portended for biodiversity. The front lines were drawn. Grassroots activists rapidly began learning about forest-plan participation, appeals, and litigation. As we peeled away the NFMA's onionlike layers, it seemed that the Forest Service, if its interpretation of diversity meant anything, knew very little about such matters.

The Mount Baker–Snoqualmie Forest Plan

"We must take a broad view of diversity and of the elements that should be considered in discussion of diversity." So wrote regional forester James Torrence in a memo to all Region 6 supervisors in December 1988.[52] "These include . . . viability, distribution and abundance of wildlife and plant populations; structure and composition of managed forests, . . . and the fragmentation of scarce community types." By drafting the memo he meant for his supervisors specifically to address fragmentation, genetic interchange and species movement and distribution, biological corridors, and plans for the reintroduction of species of concern. Quantitative data was to be used in biodiversity inventories. Torrence closed his remarks with a thinly veiled warning: "A clear discussion of diversity . . . will help us respond to challenges concerning the diversity requirements in the NFMA and the adequacy of our management for diversity."

Unfortunately, the memo fell on deaf ears. The Mount Baker–Snoqualmie National Forest is generally considered to be the most progressively managed of all Region 6 forests. Yet its final plan, issued a year and a half after the memo, is bereft of any substantive scientific analysis of biodiversity. It seems as if the Forest Service has made only a halfhearted attempt to refer to the extensive conservation biology literature. There is no credible discussion of fragmentation, landscape-scale diversity, wildlife corridors, the effect of roads on endangered species, indicator-species population levels and monitoring, and cumulative effects.

In the Mount Baker–Snoqualmie National Forest such concerns are important for several reasons. Because of past logging, very little old-growth forest remains across much of the Greater North Cascades; the landscape has been made over into various successional stands. Many of these new stands bear little ecological resemblance to natural forests and are now being managed for timber production. They will be cut again in one hundred years or less. But if some of this second growth is not allowed to mature into ancient forest, we will never be able to supplement what little old growth remains. With so many old-growth-dependent species at risk, habitat must be reclaimed. This is not likely to happen if we concentrate only on protecting the old growth that has so far escaped the saw. Martens and murrelets probably need more. The Forest Service wants more old growth for the mills. The Mount Baker–Snoqualmie plan calls for another 18,000 acres to be cut down by the year 2000.[53] (This acreage, however, will almost certainly be reduced as northern spotted owl protection proceeds.)

Logging has fragmented the Mount Baker–Snoqualmie forest into patches of varying sizes, shapes, and configurations. Many biologists have earned their Ph.D.s documenting the ecological effects of habitat fragmentation, producing a wealth of studies.[54] But the Forest Service has been slow to incorporate this literature into the forest plan, saying that few studies are specific to Mount Baker–Snoqualmie. Peter Morrison of the Wilderness Society has completed a study of how much old-growth remains on all the national forests in the Pacific Northwest. His data from Mount Baker–Snoqualmie shows that 24 percent of the remaining old-growth is in patches of 400 acres or less.[55] Forest fragments this small may reduce the value of the patch as habitat for interior-dwelling species.

There is consensus among biologists that wide-ranging species need large blocks of habitat to maintain viable populations. Most of the MIS in the Mount Baker–Snoqualmie forest depend on low- to mid-elevation vegetation types. But the majority of lands already protected in the forest are at higher elevations. The plan does not adequately differentiate between species' vital needs for preferred habitat and the paucity of low-elevation old-growth.

Logging and logging roads are the main cause of fragmentation in the forest. The ecological effects of road building are also extensively documented in the literature.[56] Yet there is little reference to these studies in the forest plan. The Forest Service mentions that roads may profoundly influence wildlife, but it does not suggest how disruptions should be addressed. Elk, small mammals, and even reptiles, amphibians, and invertebrates may be threatened by roads. For grizzlies and wolves, however, road density could mean the difference between life and local extinction.

There are but a handful of studies detailing landscape-level effects in the Greater North Cascades region.[57] Yet we have fragmented Northwestern forests and watched wolves, grizzlies, northern spotted owls, and other "landscape" species dwindle to a ghostlike few. Is there a direct connection? Conservation biology theory and studies done elsewhere suggest that there is. But without landscape-scale research, there is little empirical proof of a causal relationship in the region between habitat fragmentation and species loss.

The NFMA regulations were written to help the Forest Service to efficiently monitor actual population trends of key species. The Mount Baker–Snoqualmie final plan, however, chooses to track habitat capability in place of an actual census of each MIS.[58] With this decision, forest planners put a final twist on a concept that ecologists are still debating and that the NFMA committee of scientists did a less than perfect job of translating into law in the first place. For example, according to the final plan there could be over 4,600 pine martens (an MIS) in the Mount Baker–Snoqualmie forest simply because there are 634,500 acres of suitable habitat. The Greater Ecosystem Alliance's appeal of the plan puts this into perspective: "How will we know the effects of fragmentation, edge, roads, nonexistent or dysfunctional corridors, small 'island' reserves, or other factors on the pine marten (and those dozens of other species for which it is supposed to indicate health) if the habitat quantity, not marten quantity, is tallied? The whole intent of indicator species is to track the relationship of these two variables, not to assume it."[59] As long as the Forest Service misapplies biology, the MIS regulation holds little hope of protecting biodiversity.

If the Mount Baker–Snoqualmie forest plan was an isolated case of the misappropriation of conservation biology, the Forest Service might be excused. In fact, it is one of the better plans produced under the NFMA. The plan reduces the timber cut by over 50 percent from recent levels. The Forest Service mentions landscape mosaics, fragmentation, and other conservation biology concepts, even if no substantive use of them is made. But the Mount Baker–Snoqualmie plan is a failure. Instead of protecting biodiversity, it serves as a yardstick to measure how far away from sustainability mainstream management has veered. In the Greater North Cascades, as elsewhere, solving the biodiversity crisis will demand much more.

Challenging the Forest Service in Wisconsin

In Wisconsin, the biodiversity crisis is about to stand trial. As we have seen, the NFMA allows for administrative appeal of forest plans, and both environmental groups and the timber industry have availed themselves of the opportunity. If appeals are denied by the chief of the Forest Service and mediation bears no fruit, the appellants may bring suit in court. On the Chequamegon and Nicolet National Forests in Wisconsin, the first legal test of the NFMA's diversity provisions has been brought by a coalition of biologists and environmentalists.

No conservation biologist would ever describe the Chequamegon and Nicolet forest plans as satisfactory. And well they should know. The 1986 appeal of the plans and the current court case are based on a thorough ecological analysis by William Alverson, Don Waller, and Stephen Solheim of the University of Wisconsin at Madison. These scientists scrutinized the plans with biodiversity protection foremost in mind. The results formed the basis of a 140-page appeal by the Wisconsin Forest Conservation Task Force and the Sierra Club.[60] The appeal was also reviewed for accuracy by many eminent conservation biologists, including Paul Ehrlich, Michael Soulé, E. O. Wilson, Larry Harris, Dan Janzen, and Peter Raven, among others. What did these biologists discover?

Alverson, Waller, and Solheim found little understanding of biodiversity in the Forest Service's planning. In the Chequamegon and Nicolet plans, instead of treating *bio*diversity, the Forest Service looked at diversity defining it in the manner of foresters and landscape architects. Instead of plants and animals the agency spoke of tree-age class diversity and visual heterogeneity. Of course, the vital habitat needs of landscape-dwelling species have little to do with the percentage of young trees in a clearcut or how a cut-over area is screened from a tourist route. The plan does not describe the past, depict the present, or characterize the future of biodiversity in the two forests. Nor is there any discussion of habitat fragmentation, the future distribution of scarce ecosystem types, or the effect of edges on the remnant patchwork of undisturbed forests.

The planners' treatment of viable population is full of errors. Population estimates were derived from a selective reading of Michael Soulé's 1980 article describing the deleterious effects of inbreeding in small populations.[61] Soulé clearly points out that populations of fifty individuals or less may lose up to 25 percent of their genetic variation over ten to twenty generations. Yet the plans use this number as "consistent with short-term fitness."[62] The Chequamegon plan then goes on to estimate that a viable population of gray wolves in the forest would only need to be twenty animals!

The planners misconstrued the MIS concept as well. Instead of selecting species for each of the twenty-five ecosystem types in the Chequamegon forest, the Forest Service telescoped several distinct communities into ten "indicator communities." The majority of MISs then chosen were those resilient to human activity. Only certain types of animals were selected—no plants, invertebrates, or amphibians—including a stocked, nonnative fish. To complete this mockery, habitat capability was substituted for actual census counts of indicator species, just as in the Mount Baker–Snoqualmie plan.

Ninety-six percent of the Chequamegon National Forest would be devoted to timber production, the remaining area being managed for recreation, wildlife, and wilderness.[63] The confines of such land allocations and the poor understanding of conservation biology in the plans would end up placing native species at risk.

As an alternative, the team of biologists suggested that 17 percent of the Chequamegon be managed as "diversity-maintenance areas." Here, native forests would be allowed to reach maturity free from logging and road building, and scientists could conduct landscape-scale research.

The protection of these large blocks of forests would not lower logging levels.

The forest supervisor accepted the new idea but was overruled by the regional office, so the appeal proceeded. In January 1990 the Forest Service chief denied the appeal with no reference to the scientific arguments that had been raised. The appellants then filed suit in U.S. District Court over the Nicolet plan, doing the same with the Chequamegon plan in October. Walter Kuhlmann, attorney for the plaintiffs, has summarized his clients' main argument: "The Forest Service has continued to ignore the relationship between its prescriptions for manipulating and fragmenting the landscape and the work of scientists who are studying the ecological effects of landscape disturbance."[64]

The pending suits provide a sound scientific refutation of the Forest Service's interpretation of diversity. But biology does not necessarily translate directly into legal argument. Poor science is not automatic proof of illegality under the law. Though it appears that the Forest Service has not followed through on the NFMA diversity provisions, the law itself is quite vague and subject to interpretation. Only with knowledge of the overall context of the NFMA and the legislative history of the act can the plaintiffs claim that Congress meant to protect biodiversity across the forest landscape. Deciding whether such an argument has merit is the work of the federal judiciary. To settle the debate the courts require legal, not scientific, evidence that a law has been misconstrued.

Walter Kuhlmann has based the Nicolet and Chequamegon suits on both the NEPA and the NFMA. The NEPA mandates that the Forest Service use information on the "components, structures, and functioning of affected ecosystems," and that significant outcomes of various planning alternatives be analyzed.[65] The lack of discussion of fragmentation, change in vegetation communities over time, or movement corridors for wildlife opens the Forest Service to legal challenge. The NFMA requires the use of "biological science" as part of interdisciplinary planning and directs the agency to outline the major processes at play in the forest. No plan so bereft of conservation biology, argue the environmental groups, can be said to have satisfied these requirements. The NFMA regulations instruct the Forest Service to discuss how and to what degree each planning alternative will affect diversity in the future. Yet Kuhlmann, after four years of legal maneuvering, finds himself in court with an agency that "it is generally safe to predict . . . does not have a clue as to how each management practice or planning alternative affects biological processes in the forest."[66]

The Forest Service may be in the dark about how to manage for biodiversity in the Chequamegon and Nicolet national forests. The agency may be unable to differentiate forest ecosystems from tree farms in any of the national forests. It makes little difference, until the courts put their judicial stamp on the NFMA. And there are at least two reasons why the judiciary may side with the agency.

First, courts do not feel comfortable arbitrating disputes over scientific data and methodology. Suits based on the minutiae of cutting-edge biology do not translate well into the equally arcane language of the law. Such cases are usually found in the agency's favor. Second, in reviewing complex laws like the NFMA, many courts have decided cases based on an "abuse of discretion" standard.[67] If it cannot be clearly shown that the Forest Service has gone beyond the letter of the law in exercising professional judgment, then the suit is dropped. This reasoning depends on how professional judgment is defined and the clarity of the law in question. Opinions about Forest Service standards aside, it is helpful to remember here that the NFMA is not a model of legislative clarity.

The Wisconsin lawsuits will undoubtedly set valuable precedents on biodiversity in forest planning. The Forest Service maintains that the diversity provisions are "procedural," not substantive, and that it has no legal responsibility to manage for "any specified level of abundance or distribution of particular plant and animal communities."[68] Thus the conflict is drawn and the agency is headed for a legal showdown with the NFMA, environmental groups, biologists, and all those who wish to speak for natural ecosystems.

Forest Plans and the Courts

As judicial review looms, the NFMA planning process has come close to a standstill. It has been fifteen years since the act's passage and still, as of early 1992, not all of the forest plans are final. The Mount Baker–Snoqualmie, Chequamegon, and Nicolet plans represent a range of quality in the Forest Service's interpretation of diversity, yet even the best of these fall short of the mark. Judged by any standard, Forest Service management is still, top down, timber heavy. Though Congress wished to reduce the impact of clearcutting with the NFMA, it is still the most widely used logging technique. In dollars, 66 percent of the Forest Service's systemwide budget is devoted to timber.[69] Is this the balance Congress was looking for in 1976?

The implementation of any law is as much a matter of timing as it is of particular restrictions and reforms. Judgments against the Forest Service may be premature—the NFMA cannot be said to be fully operational until the final plans are settled and directing on-the-ground management. Many forests' cut levels, after appeals and lawsuits, will be lower than before. The threat of suit has reduced logging, but it has also spawned a politics of delay by the Forest Service. Here, we approach the limits of what any single law can accomplish as part of a change in land-management attitudes.

The NFMA prescribes business as usual for the national forests until the final plans are approved. Logging, road building, and other development activities, unless they are constrained by other laws, may continue at pre–NFMA levels. It would be to the advantage of any interest group that felt threatened by the NFMA to drag out planning for as long as possible. In 1982, the Forest Service hierarchy in Washington completely overhauled administrative forest-planning rules, sending many plans back to the drawing board. In Region 6, several completed draft plans were taken from the printer to be reworked. All the while, old-growth forestland continued to fall.

The act itself is partly responsible for planning delays. Requirements for the use of "best available data" almost guarantee that any given plan will be out of date by the time it is published. It has taken an average of about five years to write a plan, and the Forest Service had poor data to begin with. This has fueled more delays. The D.C. office has also played a role in this feedback loop. If a plan is judged unacceptable, Washington remands it to the forest from whence it came. Using this strategy, the Forest Service avoided going to court over a plan until 1988. The agency knows that once the judiciary becomes involved, managerial discretion may well be modified forever. "The introduction of non-traditional uses, such as preserving biotic diversity," says political economist Randy Simmons, "involves political conflict. Those interest groups with the most political power, best organization, and the most influence usually rule in such situations, regardless of the effects on the nation's well-being. What matters in the political system is who benefits and who pays, not what is best for people or the land."[70]

Though the courts will have the final say, there will be few clear winners and losers with the NFMA. Independent forestry consultant Randal O'Toole believes that the law's very comprehensiveness "makes it too complex for anyone to understand."[71] Certainly there has been no overnight coup on behalf of species and ecosystems. The debate between the agency and environmental groups over whether the NFMA mandates process or product is mirrored within the ranks of biodiversity supporters. Some say that the law cannot hope to work, while others hold out for the future of the forest plans.

The upshot of all this is that while Congress did not understand the importance of biodiversity, and was bogged down in compromise when the law was pasted together, it still wrote legislation that can make a difference. For concerned citizens, the NFMA means that they are no longer shut out of forest planning by the annual allowable cut and the Forest Service's imperious attitude. Planning has involved people in the battle for their forests like never before: Adopt a Forest watchdog groups have sprung up by the score, and the NFMA has provided the legal means with which to challenge timber sales and roads. Citizens have had no choice. For anyone who understands the importance of biodiversity and has paged through a forest plan, it is obvious that the rate of human-generated habitat destruction is greater than the rate of change within the Forest Service.

No entrenched bureaucracy can change overnight. But one wonders when the Forest Service will support biodiversity protection instead of reacting against it. In early 1991, the agency proposed to revise its responsibilities under the NFMA viability-population regulations. The agency felt that if a species became listed under the Endangered Species Act, NFMA viability requirements would be redundant and therefore best dropped. It seemed to matter little that Congress, in passing the NFMA, gave no indication that one law was meant to supplant the other. In 1976 Congress asked the Forest Service to bend its timber-first tradition in ways that neither lawmakers nor foresters could foresee. Today, the Forest Service continues to plead "process" and "professional discretion." That is to be expected. It will take new legislation to change old ways of thinking.

A National Biodiversity Protection Act?

"According to one of our witnesses, biological diversity is defined as 'the variety of life and its processes.' It is, in short, a fairly broad concept. To me, it means simply that life is good and the more varieties of it we have, the better. . . . But making it a part of how we operate, how we plan, where we build, what we develop, and literally how we think on a day-to-day basis is a breathtaking challenge. Fortunately our witnesses today are up to the challenges—given the stakes, they had better be—and I look forward to the recommendations they may have. . . . "[72]

With these directions, on a brisk, sunny day in November 1989, Representative Gerry Studds of Massachusetts convened a hearing on H.R. 1268, the National Biological Diversity Conservation and Environmental Research Act. I sat in the gallery of a crowded chamber of the House of Representatives' office building awaiting expert witnesses to be summoned to the stand. I strained to connect the civilized environment of Capitol Hill with the wild scent of ridgetop chaparral or sunlight glancing off the skin of a spawning Chinook salmon. It made me realize once again that what happened in this dark wood-paneled room might well mean the difference between life and death for hundreds of North American species and ecosystems. This was the third attempt at getting a national biodiversity bill passed through Congress.

Dr. Hal Salwasser, then deputy director of wildlife and fish for the Forest Service, took the stand: "Extending diversity regulations beyond recovery of federally listed species is likely to constrain short-term access to other uses of lands and waters. That is the 'no free lunch' aspect of deciding to conserve more of our biological heritage on a fixed land base. Implementing the diversity goal of the NFMA requires tough choices. . . ."[73] Salwasser did not support the draft legislation. Neither did John Turner, director of the Fish and Wildlife Service. Both men characterized their agencies as doing the utmost under existing laws. There was no need for another layer of discretion-draining restrictions.

Witnesses from the ranks of science disagreed. To a person, they supported H.R. 1268 as a boon for animals, plants, and the overall diversity of life. The bill would create a national center for biological diversity research, a federal interagency working committee whose goal would be the preparation of a national biodiversity-protection strategy, and a scientific advisory committee on biodiversity to review and coordinate research and planning among the center and the agencies. The draft would also add several layers to NEPA by mandating discussion of the impacts on biodiversity in EISs and by directing each federal agency to submit an annual report on how well biodiversity, "especially endangered biological systems," was being conserved.

These provisions were not supported by agency witnesses. The biologists, meanwhile, focused their discussion on the center. Who would oversee it—the Smithsonian Institution, the Nature Conservancy, or some other independent group? How would the $10 million in research grants be distributed? A squabble broke out between Robert Jenkins of The Nature Conservancy and George Davis of the Academy of Natural Sciences in Philadelphia. The men argued over the relative importance of taxonomic- versus ecosystem-research data gaps. Would The Nature Conservancy's Natural Heritage data base be superior to one set up by the Smithsonian? Jenkins defended his organization, while Davis made it clear he thought Nature Conservancy data was second-rate. Chairman Studds pressed the red warning button, signaling that time was up. The argument subsided and the hearing drew to a close.

In the autumn dusk, a chill wind blew in from the Potomac River as I walked to the Metro escalator. What did the people I had just heard stand for? What rang most clear to me was how each had defended his or her own organization's turf. The bureaucrats wanted nothing to do with more rules and regulations, the scientists wanted more research monies and the power to control their distribution, and if biodiversity could be protected with less of the former and more of the latter, so be it.

I did not believe that protecting special interests fit well with protecting ecosystems. The question was this: how to bring our laws and values into balance with the earth.

By mid-1992, though new bills were introduced each session, Congress still had not passed any national biodiversity legislation.

Legislation for Ancient Forests

In the politics of protecting biodiversity, setting diversity standards in 1976 may have been premature. At that time, there was little understanding of conservation biology and landscape-level connections, and no political concern over endangered ecosystems. The biodiversity crisis was building, but it had not yet begun to break. In the 1990s we are less naive, yet we continue to grope for a legal foothold against the threatened loss of bears, birds, and uncompromised ecosystems. We have not found one yet. Current bills before Congress reflect the ongoing struggle to establish biodiversity as the primary basis upon which to ground all our uses of nature.

Three draft proposals show that conservation biology is percolating into legislative language. These same bills also show that, because there is no popular consensus and so much of the economic status quo is at stake, Congress is not yet capable of voting for nature's bottom-line needs.

In addition to H.R. 1268, two bills were introduced in 1990 and one in 1991 that purported to carry the flag of ecosystem-scale protection. (Neither 1990 bill passed; both were reintroduced in 1991 with minor revisions.) The bills, which attempt to resolve the northern spotted owl crisis, label themselves solutions to the ancient forests job loss controversy in the Pacific Northwest. Scientific uncertainty over the ecological stakes in the Northwest is mostly due to data gaps. Job-loss estimates are another matter. Depending on the source, anywhere from a few thousand to 100,000 jobs are at risk. Congress has had to sort through the timber industry's (and the Northwest delegation's) claims that ecosystem protection would be the ruination of the region's economy. The truth is more complex. A 1988 Oregon Department of Human Resources report and the Fish and Wildlife Service's 1991 critical-habitat economic analysis both point to more jobs being lost because of industry automation and the sale of unprocessed logs overseas than because of environmental set-asides.[74] A closely reasoned estimate is that between 4,000 and 16,000 jobs, or about 3 to 15 percent of regional timber-industry employment, would be lost as a result of "very high" ecosystem-scale protection.[75] This is the unavoidable consequence of cutting down over 90 percent of what was once the most magnificent of the planet's forests—an inheritance of ignorance and greed. The struggle within Congress today is over reconciling new scientific knowledge justifying ecosystem protection with economic business as usual.

Representative Jim Jontz (D-Indiana), who introduced the Ancient Forest Protection Act of 1991 (H.R. 842), would solve these problems by placing the "highest value [on] intact natural ecosystems."[76] The bill would establish a national ancient-forest reserve system based on a year-long scientific study of every old-growth acre that remains unprotected on the federal lands in the Northwest. The ecological standards of the H.R. 842 are laudable—a committee of independent scientists would follow the old-growth definitions of Jerry Franklin's Forest Service task force and other research groups and design the system using the landscape-scale "biological and ecological requirements" of ancient forests. A logging moratorium would be in effect during the study period. A weak point in the bill is that it fails to provide for economic aid to displaced workers. The bill has attracted over one hundred cosponsors, but it has gone nowhere near the floor of the House. There is too much "ecosystem" and not enough "economics" in it for Congress.

In an attempt to break this impasse, Representative Bruce Vento (D-Minnesota), the powerful chair of the House Subcommittee on National Parks and Public Lands, offered an alternative. The goal of the Ancient Forest Act of 1991 (H.R. 1590) would be protection of 50 percent of remaining old-growth ecosystems. This reliance on gross percentages of land instead of ecological criteria is a weakness of Vento's bill. Instead of evaluating all unprotected old growth, this legislation would attempt to do the impossible—fix a broken system without using all the parts. Vento claims that the bill is firmly grounded in "the latest scientific information" and "advances an ecosystem approach," but this is not true.[77] Like the Jontz bill, H.R. 1590 would create a scientific committee appointed by the president to study some 6 million acres of forest over a three-year period. But the committee would only have advisory power—the secretaries of agriculture and the interior would make final decisions. And not all old-growth would be off-limits to logging during the study. In fact, the bill would mandate a cut level of some 2.6 billion board feet, further fragmenting the forests under study. Vento's bill does, however, provide for economic compensation to timber-industry workers and communities.

Vento first introduced his bill in 1990 in an attempt to push the Bush administration toward action on northern spotted owl protection and the timber economy of Region 6. When the owl was listed under the Endangered Species Act as threatened in June 1990, Bush promised to find a solution by September of that year. It was not forthcoming. While Vento prodded Bush, he also undercut Jontz, a newly elected representative with no seniority. In October, having split environmental groups between the two bills and recognizing that passage of any ancient-forest bill in 1990 was wishful thinking, Vento withdrew his bill and promised to be back next year.

The same legislative maneuvering played out in 1991 with the addition of four new House bills, mostly sponsored by various members from the Northwest delegation. All but one of these efforts was supported by the timber industry, and all were attempts to increase logging while limiting environmentalists' legal right to appeal and sue. In fact, one of these drafts, H.R. 2463, the Forests and Families Protection Act sponsored by Jerry Huckaby (D-Louisiana), would actually open national parks and wildernesses to commercial logging.

A bill sponsored by Senator Brock Adams (D-Washington), S. 1536, was different. Entitled the Pacific Northwest Forest Community Recovery and Ecosystem Conservation Act of 1991, it improved on the protection provisions of the Jontz draft while offering better economic aid than Vento's. The language of the bill made it clear that Adams had grasped a key point about the human economy being wedded to nature's: "An ecosystem-based approach to management of Federal forest lands in the Pacific Northwest is necessary for the production of timber and wood products and the employment associated with such production on a *long-term sustainable basis*; and . . . is necessary to *protect and restore* anadromous fish, riparian corridors, old growth forests, and other communities of plants and animals associated with old growth forests" (emphasis added).[78]

The economic provisions of S. 1536 were generous. Monies would be provided for unemployment benefits, job search, worker retraining and relocation, timber community economic-diversity studies, and tax incentives to reduce log exports and increase reforestation. The ecosystem-protection provisions also showed an unusual appreciation of conservation biology. The bill addressed an ecologically appropriate spatial scale: It would immediately establish an ecosystem-natural-areas network on federal lands on both sides of the Cascades and down into northern California. The traditional "crown jewels" approach to protecting nature was replaced with a comprehensive system of old-growth and associated forests managed for protection and restoration of "all natural ecological elements, functions, and successional processes." To buffer the ecosystem natural areas, a system of watershed study areas would also be designated. These lands would be fully protected for a three-year period while a scientific forest-ecosystem advisory committee studied them and made recommendations to Congress. The advisory committee would also study Forest Service and Bureau of Land Management planning efforts with the goals of conservation biology in mind: restoration of ecosystem patterns and processes, preservation of native species' genetic integrity, restoration of instream flows, biological corridors, and elimination of roads.

And there was more. The Forest Service's own research branch was directed to focus upon implementing the ecosystem-management goals of the act by studying natural disturbances, monitoring forest plan standards, and identifying which nonfederal lands might be acquired to facilitate ecosystem protection. All this research would also be subject to outside scientific peer review.

If passed, the Adams bill would serve as a model for ecosystem law across the country. But the bill's substantive standards are not likely to survive the politics of compromise. As Congress attempts to juggle logging levels, economic restructuring, judicial-review limits, and ecosystem protection, priorities will probably fall out following the order of the above list. But after years of wishing the problem away, there is no longer room for maintaining the current scale of the timber economy *and* protecting species and their habitat. John Beuter, assistant secretary of agriculture in charge of the Forest Service, made this plain in a letter to Congress. Beuter intimated that the agency simply could not cut the amount of timber specified by Congress without lawmakers "insulating the program from administrative appeals and judicial review. . . . "[79] Beuter was not describing the new provisions of the Adams bill. He was referring to the current laws on the land: the Endangered Species Act, the NFMA, and NEPA.

The Future of Legislative Reform

"Once again, oh great spirit of earth and sky, we your children are gathered unto you." These stirring words were not what one expected to hear at a Lutheran church summer camp in the heart of the Oregon Cascades. The National Conference on Forest Service Reform had little to do with religion. Over a three-day weekend in August 1989, 150 grass-roots activists, lobbyists, Forest Service workers, biologists, and even a lone congressman met to draft legislation for future forest reform.

This was lawmaking from the ground up. Throughout the conference small groups labored on a broad range of issues: timber cutting, road construction, wilderness, recreation, budgeting and appropriations, and more. The draft National Forest Reform Act had taken shape by Sunday afternoon. It did not look like anything that had ever been drafted on Capitol Hill. If biodiversity would not be served from Washington, D.C., then it would sprout green as grass from people who could still dream, care, and act:

> The task force on endangered species finds and declares that . . . the purpose of ecosystem protection is to provide a means to protect distinct varieties of native communities of plant, fish, and wildlife species which are in danger of or threatened by extinction, and to provide a program for the conservation of such ecosystems. In order

> to protect biological diversity it is necessary to manage at the landscape, or ecosystem level, rather than individual species. All federal departments and agencies shall seek to conserve ecosystems and shall use their authorities in furtherance of ecosystem protection.[80]

The conference's task force on endangered species did not flinch before these goals. It declared a moratorium on all logging and road building in national forests while a systemwide biodiversity study was conducted. Ecosystem-protection areas would then be established based upon survey results. The protected areas would be connected across the regional landscape and surrounded by areas of low-intensity logging. These, in turn, would fit into a matrix of mixed-use lands. Two overarching principles girded the task force report: Ecosystems would be managed to protect and rehabilitate biodiversity and ecological processes (structural and functional), returning them to their original integrity, and forests would be managed on the basis of current knowledge of long-term evolutionary requirements for viable populations of all species of concern (that is, 100- to 200-year plans).

This was the impact of the biodiversity crisis on the environmental grassroots, and it went far beyond the fifty-year plans the Forest Service had struggled to deliver. Activists were determined to use legal reform to challenge the collective view of forestland as resource. They were not afraid to ask, What is the bottom line? To the people gathering deep in the mountains that weekend, the bottom line was not a shrunken timber economy. Nor was it the protection of inviolate islands of wilderness. It was nothing less than biodiversity, nature's bottom line, the continuation of the evolutionary pathways writ into genes, species, ecosystems, and landscapes. It was the truth of the way the natural world works.

The ecosystem provisions of the draft reform act were beyond the ken of current politics and management. But Andy Stahl of the Sierra Club Legal Defense Fund brought perspective to the gathering: "Good teachers do not compromise on the truth."

The conferees were politically savvy. Veterans of the various wilderness wars, they knew that laws are limited—no rule can stand without broad social consensus. A law acts more like a spark than a fire. The activists glimpsed that the NFMA forest plans, appeals, and lawsuits were pushing the Forest Service toward change. And by the developing yardstick of biodiversity protection, they were beginning to sense the reach of their own political agenda: The world required more than hemmed-in fragments of wilderness and parks. Biodiversity was not just another multiple-use concept, and protecting the national forests was just the beginning of a much larger task. Arrayed against this view were the timber companies, real-estate developers, agency bureaucrats, pork-barrel politicians, and a citizenry whose ecological education never happened. Against the weight of centuries of resourcism, the activists put their faith in outlining an ecological basis for forest management. In a world balanced on the brink of change, this was the best they could hope for.

The National Forest Reform Act draft is proof that biodiversity reform is moving through legal channels, narrow though they may be. The draft has racheted us one more notch toward what law professor Robert Keiter has labeled "a common law of ecosystem management."[81] This emerging body of law, now only a rough conglomerate of endangered-species regulations, EIS processes, and vague "diversity" mandates, will eventually have to mature into an endangered-ecosystem act and other comprehensive legislation at local, regional, and national scales if there is to be any future for ecosystems. Regardless of the final shape of these laws, they must address three general concerns. First, science (and scientists) must become partners with politicians in land-management lawmaking. An advisory role is established for scientists in both the Jontz and Vento bills, but nowhere do scientists have decision making authority. (This raises deeper questions about the role of scientists and scientific information in a democratic society.) Second is the problem of the uncertainty implicit in using scientific data to produce policy. There will always be a gap between "reasonably foreseeable" and "certain." Scientists speak in the language of the former, while politicians prefer the latter. The third concern involves money. In 1987 the Fish and Wildlife Service was asked 13,824 times to consult with other federal agencies on issues of endangered species, an average 40 times per day.[82] But Congress did not provide the funds for the agency to do so adequately. Biodiversity reform will not be affordable if we attempt to fund Star Wars laser systems and Stealth bombers at the same time.

Laws may not be sufficient to effect profound change, but they do cast an educational light across the landscape. We may judge our progress in moving from isolated parks and reserves to wildlife preservation to ecosystem protection by each stage's ecological efficacy. The Greater Yellowstone fires of 1988 burned away administrative boundaries and showed politicians a minimum dynamic area in action. In the North Cascades, 35 percent of the region is designated wilderness, yet many species are in jeopardy. It matters where wildlands remain and how they are connected, not how many acres are protected. Before 1990, wildlife law treated animals as if they were stationary plants, but in September the U.S. Court of Appeals barred the Forest Service from allowing logging on the last roadless link between two wilderness areas in the Klamath National Forest in northern California.[83] It was the first explicit legal recognition of the reality of wildlife movement across ecosystems, and it made the front page of the San Francisco newspapers.

CHAPTER FIVE

THE LANDSCAPE OF MANAGEMENT

No expert knows everything about every place, nor even everything about any place. If one's knowledge of one's whereabouts is insufficient, if one's judgment is unsound, then expert advice is of little use.

WENDELL BERRY,
Damage

Having studied the Greater North Cascades for only eight years, I felt somewhat uncomfortable in my role as an invited guest speaker at the 1987 U.S.-Canadian Transboundary Resource Conference at Western Washington University in Bellingham, Washington. I had been invited to give a summary of my studies of the potential for an ecosystemwide management approach in the region, and I wanted to be provocative while making some practical suggestions for better cooperation. But I was unsure how to capture the attention of my audience. As I walked under the rain-soaked maples toward the conference room, images of mountains flashed through my mind: white ghost flowers of early spring, summer mist fingers wrapping around high-country firs, autumn larches in still life above Cutthroat Pass, the bruised hide of a spawned-out salmon, wet blankets of new snow outlining clearcuts higher up in the foothills. I considered the links between these images and the uncertain future of the Cascades north and south of Koma Kulshan.

Twenty-seven forest and park managers, agency bureaucrats, academics, and activists gathered in the conference room. Jim Torrence, the Region 6 forester, could not recall a time when a topic other than timber had been the focus of a meeting. Paul Pritchard, president of the National Parks and Conservation Association, did not remember being in the same room with a former Park Service director, a Park Service regional director, and a regional forester, unless it was at an awards banquet in Washington, D.C. British Columbia parks director Jake Masselink had never attended a meeting with his U.S. counterparts. Few had ever had the occasion to share their North Cascades experiences with peers.

Today promised to be different. The group was meeting on the neutral ground of the university to discuss the fate of the wildest mountains in the Pacific Northwest. Many of us wondered what management policies might emerge if, even for a single day, we could erase administrative boundaries and consider the North Cascades as a whole.

"Is it possible for us to cooperate on regional matters?" began host John Miles, dean of Huxley College of Environmental Studies.

"I've come here to listen," replied Jim Torrence. "But we already share law enforcement and fire suppression with the Park Service." It was clear to most conferees, however, that fighting fires and writing traffic tickets was not regional management. What about the impact of the forest plans on North Cascades National Park? Who would take responsibility for all the clearcuts planned hard against the park's boundaries? Who was to be accountable for the movement of deer, elk, and cougar between multiple-use and protected lands?

Jake Masselink noted that this was the first time in nineteen years that Canadian managers had been invited to discuss transboundary issues with the U.S. agencies. "The North Cascades do not all of a sudden disappear across the forty-ninth parallel," he reminded the group.

"But environmental protection laws do disappear," shot back Peter McAllister of the British Columbia Sierra Club. "We have no ecosystem planning in Canada, nor do we have laws like the NEPA or the Wilderness Act. Public participation in resource planning is almost nonexistent."

"Here in the States we have too many plans," offered Charlie Raines of the Sierra Club office in Seattle. "We can't comment on three 500-page forest plans all at once with only volunteer labor. Why should citizens bear the brunt of keeping the Forest Service on track?"

Later in the day it would be my turn to speak. I considered my prepared remarks and another image from the past welled up, a passage from a paper written by a United Nations consultant decades ago: "We have come to a period in human history when there is a great need for . . . landscape or regional planning. . . . a multidisciplinary, multiagency, public-private joint planning that can only stem from the existence and nature of natural . . . ecosystems."[1] But the group would not be swayed with talk of dissolving political boundaries or improving forest plans to provide wildlife corridors: No one in the room grounded their management in the nature of natural ecosystems. They based their decisions on socially constructed images of "natural resources" mediated by administrative politics within large public and private bureaucracies. Knowledge of viable populations, habitat fragmentation, buffer zones, and natural disturbances was not widespread here. Politics, not ecology, was our common currency.

As one who had some insight into both of these worlds, how could I outline common ground between science and expediency?

What Is a Resource?

The North Cascades Transboundary Conference was not my first brush with the federal land-management agencies. That had come a decade earlier. It took only a few summers of seasonal labor with the Forest Service in the 1970s for me to grasp the government's style of managing resources. We would burn slash in old clearcuts when there were no wildfires to contain, plant genetically selected nursery stock in clearcut units, and chip brush along roadsides to retain "visual diversity." Working for the Park Service, I sold hundreds of postcards to tourists after presenting a ten-minute slide orientation on the scenic highlights of the park. Several seasons as a backcountry ranger granted me the privilege of living in several national parks. I would listen to walkie-talkie static fifteen miles from the nearest road while checking hiking permits.

Stacking sticks and writing permits helped to put me through college; it also encouraged me to think critically about our collective management of nature. Why was it important for the Forest Service to maintain a scenic corridor of old growth between highways and clearcuts? What was the point of viewing ancient forest as so many board-feet? Why were most wilderness areas located in the high country? How did it come to pass that some level of development seemed always to be the "preferred alternative" in management plans? Who determined standards of use, enjoyment, and preservation? And why were the agencies continually feuding with each other? In short, why were the national forests turning into tree farms and the national parks becoming magnets for "consumer-style recreation"?

None of these questions was broached at the Transboundary Conference. The agenda didn't address the ideology of agencies. We had gathered to assess the justification for regional approaches to managing the North Cascades, and it was clear that some participants felt current efforts were sufficient. It was as if several ostriches, each with its head buried in sand on a different side of the same mountain, had momentarily stood up to take stock of its surroundings. Having done so, the majority were quite ready to resume their accustomed pose.

But, at the time of this conference, the biodiversity crisis was shaking up the status quo, and management styles were shifting. By the 1980s it was becoming obvious that agency ostriches would have to adopt a new stance. Citizens in increasing numbers were adding their voices to the land-management debate. Grassroots groups, formed to protect local forests, watersheds, mountains, and threatened species, sprouted like mushrooms after fall rains. Arguments at environmental hearings became increasingly volatile. Pressure was mounting, and by the time of the Bellingham conference even a handful of managers were beginning to call for something called ecosystem management. But if nature was structured in ecosystems and the federal agencies were managing nature, wasn't ecosystem management already being practiced?

It may be surprising, but land managers don't fit their practices to nature any better than Congress crafts laws to fit the land. Clearcuts, inadequate reserves, scientific ignorance, and meager funding exist because most people are more interested in "denaturing" nature than in molding themselves to the world they are part of. "As we think," biologist Chris Maser has remarked, "so we manage."[2] It is impossible to understand why the current landscape of management has yielded fragmented forests and feuding agencies without first probing the roots of entrenched Western attitudes toward nature.[3] Until we have thought through our concepts of nature, resources, and management, we have no hope of changing the status quo.

History may help us peer through the veil of our environmental values. A tale of two agencies, the Forest Service and the Park Service, is a good place to begin.

The Origins of Public Land Management and the Rise of the Forest Service

Americans today take public lands for granted. But neither national parks nor national forests existed before the latter nineteenth century. In fact, there was no public domain of any kind until 1781. At that time, the obscure, overlapping, and contentious Western land claims of the original thirteen colonies were settled in favor of national ownership. Between 1781 and 1802 the public lands grew to 237 million acres.[4] This was only the beginning of westward expansion.

By 1867, when the purchase of Alaska was negotiated, the public domain had grown to 1.84 billion acres through a combination of outright purchase (the Louisiana and Florida purchases, the Oregon Compromise, and others) and war (the Mexican Cession). Almost 80 percent of the country was now in federal ownership, and land became the richest resource of the newly expanded nation. With no management policy and less ability to enforce one, Congress began to give land away to new states, settlers, soldiers, and speculators in hopes of encouraging investment and settlement.

Though some envisioned the New World as an earthly paradise, many of the first Euro-American settlers came loaded down with cultural baggage, regarding nature as something to be feared and conquered. Historian Roderick Nash has shown that, for the pioneer, "the transformation of wilderness into civilization was the reward for his sacrifices, the definition of his achievement, and the source of his pride."[5] Whether for salvation, pride, or profit, the new settler's image of nature collided with the incredible diversity of the natural ecosystems of the United States. The country's forests, minerals, and grasslands were perceived as inexhaustible. America's pioneer land-use policy was based on what historian Craig Allin has called "the economics of superabundance": "There were always more resources; there was never enough labor. It is only logical that families moved on to greener pastures after a few years of farming depleted the soil. New land remained cheaper than more intensive management. Similarly, loggers burned or left to rot many times the value of lumber that they actually marketed. With such abundance it was unprofitable to utilize any but the best trees; the rest were simply impediments to progress."[6]

The last decades of the nineteenth century were the rowdiest chapter in the history of U.S. land management. Congress passed a plethora of land-disposal laws, many of which contradicted others. None was implemented with any consistency. With lax enforcement, there was really little difference between private, state, and federal ownership in the West. Squatters camped where they wished. Timber companies cut down forests they owned and many more they did not. Ranchers "legally" claimed gargantuan spreads and then set hordes of cattle to overgrazing. There were few fences to stop them. The railroad companies may have made out best of all. In return for constructing an iron lifeline of commerce from east to west, the government granted the Northern Pacific Railroad (Burlington Northern today) 21,000 acres for each of 2,128 miles of track.[7] By 1870 Congress had given away 182 million acres to railroad companies alone, an area almost equal in extent to today's national forests.

Westward expansion and unfettered development made the United States the wealthiest country in the world. Americans mixed their labor with the land and created a high-speed economy geared toward profit. Population, production, technological innovation, and specialization all grew exponentially as the country fed its ecosystems into the mills of industrial capitalism. As the nineteenth century drew to a close, however, a few visionary Americans were beginning to see beyond the taming of the continent. Land grabbing and resource depletion could not go on forever. Federal land management in general and the national parks movement in particular grew out of an evolving sense of the need for land-use limitations. This trend was strengthened by an increasingly urban and leisured middle class that looked upon rapacious development and lack of federal oversight as undemocratic, inefficient, and against the national interest.[8] Certainly the conservation and preservation movements were born in America's first flush of material abundance. Preservation and protection, aesthetic luxuries at the beginning, were yet to become ecological necessities.

President Benjamin Harrison, with the consent of Congress, carved the first forest reserves out of the vast public domain of the western United States in 1891. Pressure had been building for lawmakers to withdraw lands to protect them from squatters, speculators, and trespassers ever since the first national parks were proclaimed in the 1870s. Even so, Congress did not act decisively—the new reserves were given no *raison d'être* or funding.[9] They were drawn on maps, but no one knew what purpose they served.

The birth of federal land management was no cut-and-dried affair.[10] There was no sudden abandonment of utilitarianism for wilderness preservation. In 1891 the majority of Americans did not consider forest reserves essential to the welfare of the nation. To some they were sacred; to others they were obstructions to progress. Though Congress had begun to reverse its land giveaway policies, it remained to be seen what would be done with the newly consolidated public domain. This was the dawn of a new era. It takes imagination for Americans of the last decade of the twentieth century to comprehend the landscape of the last decade of the nineteenth. Today, with 80 percent of the continental United States in private hands, ownership patterns have been reversed in the course of a century. One hundred years ago, Western settlement was by no means complete and there were no public-lands policy precedents. There was not even a Forest Service or a Park Service to manage federal land.

Several people, John Muir and Gifford Pinchot among them, had some ideas about managing the new public lands. Muir, the foremost proponent of national parks, wanted the forest reserves left alone, set aside for preservation in the manner of Yellowstone and Yosemite. Pinchot, on the other hand, felt they should become the nation's first forestry laboratories. Fresh from France and Germany, where he had studied silviculture, Pinchot was eager to plant European forest-management ideas in the soil of the New World. Both men became members of the 1896 National Academy of Science commission whose purpose was to recommend a forest-reserve strategy to an undecided Congress. After a summer of touring the reserves, the commission, unable to reach a consensus, sidestepped the main issue and suggested that more lands be set aside. When President Grover Cleveland complied by doubling the acreage of the existing reserves, Congress reacted strongly. With a coalition of Western congressmen adamantly opposed to any new land withdrawals leading the way, the Forest Management Act of 1897 passed by a two-to-one margin. The new law authorized the secretary of the interior, who administered the forest reserves, to sell timber and protect watersheds. The reserves were not to become preserves. Pinchot won, Muir lost, and the country had its first official land-management policy.

The argument Pinchot and Muir had over the future of the forest reserves split America's nascent conservation movement. Some followed Pinchot and favored "wise-use" of nature; others believed, with Muir, that forests were to be preserved forever.[11] That Pinchot had his way with Congress is hardly surprising given the American experience up to the 1900s. But how exactly did the first chief of the Forest Service define "wise-use"?

Pinchot did not often practice the political doublespeak so common today. Though an accomplished Washington insider, he argued his ideas with blunt and forceful clarity: "The first great fact about conservation is that it stands for development."[12] The chief forester summarized this philosophy with the slogan that conservation should provide "the greatest happiness for the greatest number over the longest period of time."[13] That his views were later codified by Congress with the passage of the Multiple-Use Sustained Yield Act in 1960, and that they still form the bedrock belief of Forest Service managers some nine decades later, is a tribute to Pinchot's shrewd vision and political acumen.

Questioning Pinchot's famous dictum, however, is a lesson in critical thinking. Who decided what happiness was, and how much was enough? Certainly not Congress—lawmakers had mentioned only timber and water in the 1897 Forest Management Act. How was satisfaction of the "greatest number" to be determined? Were certain groups automatically positioned to reap the benefits, or were forest products and profits to be distributed equally among all classes of society? What about nonhuman species' claims for happiness and habitat? How long was the "longest period of time"? Pinchot responded that resource use should first serve the needs of "people who live here now."[14] Don Snow has suggested that Pinchot's multiple-use philosophy derived as much from the politics of satisfying a multiple group of commodities interests as he built his national forest empire as from European forestry.[15] Either way, Pinchot communicated little vision of the long-term future.

Pinchot's neat sustained-yield slogan implied that choices over how to manage nature were subject to objective economic measurement. He automatically discounted all other values; standards of land management and ecological health were based on the amount of forest products and other commodities wrested from the woods. Pinchot saw the Forest Service as an alternative to the cut-and-run tactics of private timber companies, and he wanted to use scientific forestry to solve forest-management problems. But, while his European schooling taught him that forestry should be scientifically based, it did not show him that science could also contribute to the destruction of ecosystems. Several key misconceptions about nature bolstered his brand of conservation. Chief among these was the illusion that forests are stable over time.[16] In a static natural world, a forester only had to worry about the short-term replacement of a new crop of trees after harvest. This image of nature allowed Pinchot and his men to design management plans atomistically—trees were merely replaceable cogs in the machinery of nature. A manager could lay out a timber sale or prescribe a grazing plan without taking into account neighboring stands, the larger watershed, or the region. With nature stable and divisible, a forester could act with certainty: "With skillful handling and within the limitations imposed by the site itself, we can produce any combination (of uses) we want," proclaimed an early *Journal of Forestry* editorial.[17]

Even as the first forest rangers spread into the Western hinterlands, North America's forests and grasslands remained in the powerful grip of "long-held Western philosophical and religious beliefs and assumptions that humans were destined to dominate Nature and that Nature existed solely for the use and enjoyment of humans."[18] Pinchot's vision of forests, shaped by mensuration tables and cruising sticks, reflected the turn-of-the-century Progressive Era. The reforms of Progressivism offered an antidote to excessive exploitation: an expanded view of the national interest, an optimistic vision of the future, and more government control.[19] The Forest Service rose to the task, but while conservation carried the day, at root nothing really changed. With the excesses of expansion being challenged by both Pinchot and Muir, Americans had the opportunity to re-envision their role within an ecological context. What they opted for instead was Pinchot's revisionist conception of humans as wise, scientific managers of balanced resources.

Rivalry with the Park Service

To the casual observer, the national park movement appears to offer an alternative to the resource conservation philosophy of Gifford Pinchot and the Forest Service. With his deep respect for wild nature, preservationist John Muir was certainly no Pinchot.[20] But Muir, for all his popularity, was too wild, too visionary. Though influential as a writer, his ideas never did enter the political mainstream, and he remains a guiding star toward what he might have called "righteous" ecosystem management. Muir could not effectively counter the wise-use ideology of his day. Although he lobbied tirelessly for preservation in general and the national parks in particular, the Park Service today bears little of his stamp.

Like the Forest Service, the Park Service was a child of the Progressive Era. From the beginning, however, the two agencies emphasized different aspects of wise-use. While the Forest Service advocated scientific production for human benefit, the Park Service anchored itself in nineteenth-century transcendental views of nature as a place of spiritual encounter. The parks symbolized a balanced harmony between wilderness and civilization, providing respite for citizens amidst the rush of the rising American empire. Muir exhorted people to "climb the mountains and get their good tidings." While many rode the railroad to Yellowstone, most stopped short of his view of nature as the source of all value. According to the 1916 National Park Service Organic Act, parks are meant to preserve nature for future generations and provide recreation. This management concept moved well beyond Pinchot and the Forest Service by defining long-term use; still, the parks were conceived exclusively for humans—natural museums on a magnificent scale. The young country, searching for a national aesthetic that could compare favorably with Old World charm, was proud of its park showcases, not because they were wild but because they were grand.

The Park Service shared the balance-of-nature view with the Forest Service. Only in this version, instead of an endless flow of timber products, a stable nature yielded primitive vignettes for the sublime delight of tourists. Federal control was the best path to this goal. Philosopher Warwick Fox is correct when he points out that conservation and preservation spring from the same anthropocentric roots.[21] Muir and Pinchot wrangled over this very point: Do humans stand with or stand above the rest of Earth's life forms? To this day, the conflicts between the two agencies revolve not around putting humans first but rather around multiple-use versus preservation-oriented land management.

The political feuding between the two agencies is born of their different approaches to the instrumental use of wild nature. With the forest reserves firmly in his grasp after a 1905 agreement that transferred them from the Interior Department to his agency in the Agriculture Department, Pinchot lobbied against the creation of the Park Service.[22] He believed that the parks should be folded into his agency and opened for development. Powerful lobbyist though he was, Pinchot could not sway Congress and the president in his direction.

Once the Park Service was established it sought to expand, while the Forest Service attempted to protect its domain. Most of the early parks had been carved out of Forest Service lands, and this pattern continued. During the 1920s recreational use of both parks and forests skyrocketed, leading to hard-fought turf battles between the two agencies. Both believed they should have exclusive control over recreation on the public lands. By 1960, 30 percent of the parks had been created from Forest Service lands.[23]

How did the Forest Service come to value recreation as a legitimate multiple use alongside timber and grazing? Most public-land historians agree that the agency embraced wilderness as a strategy to minimize Park Service acquisitions.[24] If a given area of national forest was already administratively designated for primitive recreation, Congress would have no reason to transfer it to the Park Service. This strategy had mixed results for both agencies. Although the Park Service did expand, the Forest Service retained the majority of its scenic lands. These political battles continued through the passage of the 1964 Wilderness Act and up to today, leaving a legacy of distrust between the agencies that augments their philosophical differences.

Wise-Use Today

Park/forest, Park Service/Forest Service, federal/private, United States/ Canada, owls/jobs, wilderness/civilization, nature/people—our thinking is shaped by these deep divisions in our Western image of the world, and they affect every moment of our lives. "The Park Service stole the North Cascades from us in 1968 and I'm never going to let them forget it," raged a prominent forest supervisor to me in his private office just a week before the Transboundary Conference.

"The feds had better get their act together or the Cascade River is going to be logged to kingdom come," warned a Skagit Valley resident at a public hearing. Yet, when I queried him later about conservation easements to reduce the threat of logging on private lands, he railed against government meddling in citizen affairs.

"My park shares twenty-five miles of common boundary with British Columbia and I have no idea what is going on with wildlife populations across the border," admitted Jarvis, resource management specialist for North Cascades National Park, as we stood drinking coffee before the Transboundary Conference. As we talked, I realized that I didn't know any more than Jarvis about elk, owls, and grizzlies across the border. Later, as I followed the thread of the conferees' discussion and heard little mention of the distance between agencies, I knew that, whatever I chose to present to the group, it would have to be cast in language understandable to high-level administrators. But it would also have to tell the truth.

Under the guise of wise-use, wild nature has shrunk before civilization, walled off into parks and preserves that are slowly losing their ecological integrity. Both the preservationist vision of parks as protected paradises and the Forest Service view of multiple-use lands as a resource cornucopia have "encouraged us to believe that conservation is merely a system of trading environmental write-offs against large protected areas."[25] After 200 years of American exploitation of nature, management is much more than a question of balancing acres of parkland with forest. Most people now live in cities bereft of any but the most degraded ecosystems. Seeking solace, they drive to public lands far away in mountain and desert, passing through rural country that hides its ecological wounds from all but the trained eye. In the national forests that surround many of the parks, we are sheltered from the rape of the woods by carefully planned "viewsheds." Few explore beyond the screen of the main highway; vacation time is precious and our destination lies ahead. Upon reaching the national park, we seek the scenic vista, the awesome heights, and camp in developed campgrounds complete with trailer hookups and paved nature trails. The vast majority of park visitors never set foot in the backcountry. This is the architecture of splintered consciousness: city/country, forest/park.

The history of interagency land management has contributed to this rupture. Multiple-use management for human commodities has never been balanced on Forest Service lands. The Park Service, riding the fence between use and preservation, has too often erred on the side of the former. The agency has never acknowledged that managing for amenities serves only narrow human interests.

To begin to solve these problems we must adjust our mental boundaries, renegotiate the political borders of the landscape of management. Wendell Berry notes that "to use or not use nature is not a choice that is available to us. . . . Our choice has rather to do with how and how much to use."[26] Biologist David Hales has translated Berry's wisdom for land managers. To Hales, it is not a question of "whether the integration of parks and surrounding lands is important, but how integration should occur and to what degree."[27]

Pinchot's wise-use resourcism responded to the pillage of North America at the hands of the first generations of pioneers and captains of industry. Today the biodiversity crisis is fostering change—from resource conservation for commodities and amenities to ecosystem management for native diversity. The story of the rise of federal land management shows that the American image of nature was little altered after John Muir lost the forest reserves to wise use. It remains to be seen whether the new ecosystem view of management will have a more profound influence.

The Story of the Northern Spotted Owl

I have never seen a northern spotted owl. Newspaper photographs have shown me a small, unprepossessing bird that seems, like most owls, to be somewhat uncomfortable in the daylight world. Owls are night hunters. The genus *Strix*, to which the northern spotted owl belongs, shares silent flight with the rest of the owl clan. Unlike other birds, owls have serrated primary feathers on the leading edge of each wing. Air flow over smooth primaries produces noise, but owls fly soundlessly. Owls also have finely tuned hearing. They can accurately locate prey in the dark because their ears "separate" sounds—there is a lag in hearing time between one ear and the other. Asymmetrical ear openings give them a further advantage by registering different sound intensities up and down the ear. Special cells located in the midbrain help owls to organize these complex signals into what biologists have called a "neural map of auditory space."[28] Because owls usually inhabit a single territory for life, this map is overlaid nightly on familiar ground and a silent swoop results in supper.

Unfortunately for the northern spotted owl, its preferred habitat is almost exclusively old-growth forest. Up and down its range from British Columbia to northern California, the owl is dependent on forests that are 200 years old or more. These, of course, are the same forests that timber corporations covet, environmentalists defend, and the Forest Service must manage properly. The owl is an MIS in all of the westside national forests in the Pacific Northwest, and so its population, under the NFMA, must remain viable. Easier said than done. *Strix* has become the most celebrated and vilified endangered species since the snail darter.[29] Although the end of the spotted owl story has yet to be written, the bird has already altered the landscape of management forever.

No one knows exactly why the northern spotted owl is so closely linked to ancient forests. The owl prefers to nest in the lightning-sheared tops of live Douglas-fir, western hemlock, and western red cedar trees. In winter, the dense, multilayered old-growth canopy provides thermal cover. In summer, the cool shade cast by centuries of woody growth offsets the bird's thick plumage, which limits its ability to dissipate heat. The canopy may also shield the owl from its predators while producing the low-light conditions under which its rodent prey thrives. Nesting spotted owls have been found in stands that are not classic old-growth, especially in northern California, but these birds are the exception rather than the rule. Most spotted owl experts agree with ornithologist David Wilcove of the Wilderness Society that these "second-growth" individuals are so few as to be "simply irrelevant to the long-term survival of the species."[30]

It is hardly irrelevant that over 90 percent of the spotted owl's remaining suitable habitat is found on federal lands. With most privately owned ancient forests long gone, the pressure to log remaining national forest old-growth is barely exceeded by efforts to protect the owl. The last chance to compromise on this issue was lost at least ten years ago when the Forest Service ignored the biological handwriting on the wall and, instead, began work on the first of what has become an ongoing series of spurious northern spotted owl management plans. Taken together, these plans prove that wise-use conservation still holds a firm grip over federal land management.

At first, so little was known about northern spotted owl biology that the Forest Service could afford to fund a few studies while continuing to liquidate ancient forests. But research yielded disconcerting results. Spotted owls not only needed old-growth, they needed lots of it. Each pair of birds in northwestern California requires about 1,900 acres of uncut old-growth, while those in the Washington Cascades need at least twice this amount.[31] (The discrepancy is likely due to north-south differences in prey availability.) Demographic studies have shown many of the signs of an extinction-prone species. Only about 2,900 pairs are known to exist. The birds do not breed until they are three years old, and then their breeding rate is uncertain. Juvenile mortality is extremely high; since the 1970s, overall first-year survival has averaged only 11 percent.[32] During this time their prey base has also fluctuated widely. Overall, the population growth rate is very low.

Much of the juvenile mortality results from the young owl's inability to cross clearcuts and successfully disperse in a fragmented landscape.[33] Fledglings cannot easily negotiate logged areas and early successional stands without being picked off by predators. The great horned owl thrives in edge habitat and prefers to dine on northern spotted owls. The introduced barred owl also outcompetes the spotted owl in a patchy landscape and seems to be displacing adult pairs. Some species of owl find easier hunting in clearcuts, but all scientific evidence points toward a declining population for the northern spotted owl.

Facing these studies, the Forest Service acted predictably. The first northern spotted owl management plan was released in 1984.[34] Two hundred sixty-three pairs of owls were to be protected by a network of "spotted owl habitat areas," each 1,000 acres in size. The plan ignored the scientific evidence that owl pairs in Washington need almost four times that amount of habitat. And some of the habitat areas were placed in cutover lands that had no resident owls. Environmentalists quickly appealed the guidelines and the secretary of agriculture directed the Forest Service to prepare a supplement that addressed their concerns.

Expecting little change, the National Audubon Society asked the two most prestigious U.S. ornithological groups, the American Ornithological Union and the Cooper Ornithological Society, to convene a scientific panel to review the literature and make management recommendations. The panel released its report in 1987, concluding that at least 2,900 pairs of owls needed protection on federal lands with an average of 2,300 acres/pair in California and Oregon and 4,200 acres/pair in Washington.[35] This would ensure population viability over the next fifty years. The report established a scientifically based strategy for the northern spotted owl, and yet the trees continued to fall.

Any doubts about Forest Service intentions were dispelled with the appearance of the supplemental EIS in 1988.[36] The agency stood firm with a belligerent timber industry and the protimber Northwest congressional delegation. The EIS proposed to protect only 1,500 to 2,000 pairs of owls in habitat areas sized far below what was known to be necessary for the bird's survival. The Forest Service's projection for the owl was chilling: After 100 years, there would be only a low to moderate probability of population persistence. A low probability meant that "catastrophes [and] demographic and genetic factors are likely to cause elimination of the species from parts or all of its geographic range."[37] Not even the bureaucratese could hide the fact that this would be extinction by averting one's eyes.

As the Forest Service plan headed for a legal showdown, a separate court challenge was being launched. In 1987 a coalition of environmental groups petitioned the Fish and Wildlife Service to list the owl under the Endangered Species Act. The agency responded with a "no listing warranted at this time" ruling that was contradicted by rapidly accumulating scientific evidence. Called in to investigate, the Government Accounting Office, the watchdog arm of Congress, found that the Fish and Wildlife Service hierarchy had "substantially changed the body of scientific evidence presented in the study team's status report after it had been reviewed and adjusted by outside experts."[38] Environmental groups sued, won in court, and the spotted owl petition was sent back to the Fish and Wildlife Service for reconsideration. Soon afterwards in

Seattle, federal Judge William Dwyer issued an injunction against the Forest Service, and logging in old-growth was banned in Region 6. The owl was winning in court and the pressure was on the agencies.

The saws were not silent for long. Unwilling to wait for the owl to be listed under the Endangered Species Act while logging was held up by court injunction, Senators Mark Hatfield (R-Oregon) and Brock Adams (D-Washington) attached a rider identified only as section 318 to a Senate general appropriations bill. This passed with little opposition. A similar measure sailed through the House. Just as the Endangered Species Act and the diversity provisions of the NFMA were being forced upon the Forest Service by the courts, Hatfield and Adams performed a sleight of hand that made a mockery of both the law and the democratic process.

Section 318 had both expected and unexpected consequences. It canceled the court injunction so logging resumed, only this time under a mandated target of 3.85 bbf per year through 1990. This outrageously high cut would "stabilize" the timber industry while Congress searched for a permanent solution to the northern spotted owl crisis. Section 318 also limited judicial review of old-growth timber sales. Environmental groups could only challenge the Forest Service in court under extremely narrow circumstances. This part of Section 318 was likely unconstitutional, but, for the moment, it made little difference. As a bone to the defenders of the owl, the Hatfield-Adams rider created the Interagency Scientific Committee to Address the Conservation of the Northern Spotted Owl. Headed by the dean of Forest Service wildlife biology, Dr. Jack Ward Thomas, the committee was charged with producing a scientifically credible conservation strategy for the bird. As loggers went back to the job of reducing old-growth to so much sawdust, and Representatives Jontz and Vento began drafting their own bills to save the ancient forests, the Thomas committee huddled with conservation biologists and owl experts.

"Scientists Side with the Owl," proclaimed the front page headline of the *Seattle Post-Intelligencer* on the morning of April 5, 1990. Ever since owls and old-growth had entered public discourse, the media had portrayed the argument as owls versus jobs. Though jobs were certainly at stake, the issue was much more complex. Protecting biodiversity will never boil down to one species against an industry. The fate of the northern spotted owl is an index of the fate of old-growth forest itself. All Northwest species, humans included, depend upon clean air, fresh water, and stable soils—all ultimately have an equal stake in the ecosystem's health.

The Thomas committee report recognized these ecological facts of life from the beginning. The committee adopted a 100-year time frame and used a clutch of conservation-biology principles: (1) species that are well distributed across their geographic range are less prone to extinction than species that are confined to limited portions of their range; (2) large blocks of habitat containing multiple pairs of a species are better than small areas with only a few pairs; (3) habitat-block spacing should reflect the dispersal biology of the species at risk; (4) contiguous habitat is better than fragmented habitat; and (5) it helps dispersal when connecting corridors also have suitable habitat characteristics.[39] The Forest Service habitat-area strategy was jettisoned in favor of large-scale habitat conservation areas (HCAs) designed to protect up to twenty pairs of owls each. The areas were located close enough together to accommodate known juvenile dispersal distances, and no further logging of any kind was to be allowed within them. Most important, the report placed logging restrictions on spotted owl habitat *outside* the HCAs. This landscape-level framework would preclude the need for designated dispersal corridors as long as the committee's "50-11-40" rule was followed, where "at least 50% of the forest landbase outside of the HCAs is maintained in stands of trees with an average diameter at breast height of 11 inches and at least 40% crown closure."[40] (This would amount roughly to a forty-year-old forest.) The biologists were charged to protect the owl, and it did not matter whether the bird lived on federal, state, or private land.

Of course, land ownership did matter. About 50 percent of the HCAs were located in parks, wilderness, and other protected federal lands. The rest, in areas of mixed ownership, would also be removed from timber production, but only if the various parties signed on to the Thomas report. And not all forestland within the HCAs supported spotted owls. Given the Northwest's fragmented forests and the fact that owl habitat had been logged for a hundred years, the committee could only delineate HCAs that had 50 to 60 percent of their potential spotted owl habitat intact.[41] This was one reason for the report's long time frame: The strategy could only succeed if there was time to grow back ancient forest by allowing logged-over lands to mature. The bits and pieces of old-growth that had survived into the 1990s would not be sufficient. To protect northern spotted owls, the committee banked on owls and old-growth yet to be born.

It is one thing to hold on to something you already have, say, a diminutive bird, flesh, blood, and feathers, in your hand. It is quite another to act with the distant future in mind. Even with their conservation strategy fully implemented, the Thomas committee expected a 40 to 50 percent decrease in the current owl population.[42] With old-growth being lost outside the HCAs, owl populations would continue to decline. Inside the HCAs, populations would increase at the same rate that trees matured; owls outside the network would die as quickly as trees were cut down. This was the measure of risk in the Thomas report, and it stretched to the end of the twenty-first century.

The gamble seemed to go against the most recent population-viability model for the owl. Dr. Daniel Doak of the University of Washington concluded his spatial-demographic analysis with a simple plea: "[S]uspension of all old-growth harvesting is the only prudent course of action in the forests of the Pacific Northwest."[43] Jack Ward Thomas agreed. Before a congressional committee in Washington, D.C., Thomas testified that "most, if not all, spotted owl experts agree that cessation of all old-growth logging would be best for the owl."[44] But Thomas was no fool. He recognized that a complete moratorium would not be politically acceptable. While the report banned logging in the HCAs, the untested 50-11-40 rule on lands outside protected zones represented the compromise between science and expediency. What lay behind this experimental standard? What would be the impact of the rule as forests matured over twenty years? Fifty years? One hundred years? Thomas made it very clear that if all landowners did not follow 50-11-40, the line between viability and extinction would disappear. But this meant that a federal interagency plan would have to be enforced on state and private lands. The Forest Service had been directed by section 318 to consider the Thomas report's conclusions in any old-growth management decisions. Did the government expect to tell property owners how and when they could cut the most valuable forests in the United States? Once again, conservation biology was pushing land managers, public and private alike, toward a new future.

"The biological evidence says that the northern spotted owl is in trouble. We will not, and by law cannot, ignore that evidence." Thus spoke John Turner, director of the Fish and Wildlife Service, in June 1990 as he officially declared the owl threatened under the Endangered Species Act.[45] Turner's pronouncement, along with the months-old Thomas report, ignited a political firestorm.

"I'm not going to allow a noose to be placed around the neck of the Northwest economy and then let Jack Ward Thomas be judge, jury, and hangman," ranted Bureau of Land Management director Cy Jamison.[46]

"I found an owl last week in Oregon and it wasn't in old-growth timber," Manuel Lujan, secretary of the interior, recounted. "So it showed me that the owls can live outside of old-growth areas."[47]

Senator Bob Packwood and Representative Denny Smith, both Republicans from Oregon, had another idea: captive breeding of the owl in zoos. "We're doing it with the California condor," said Packwood, "why not with the owl?"[48]

President Bush sought the middle ground. "I don't think we can accept an answer that's going to paralyze employment in the Northwest," he cautioned.[49] Bush caught the political football and immediately blamed the Thomas committee and an indecisive Congress. He then appointed (yet) another interagency task force, this time headed by Secretary of Agriculture Clayton Yeutter, to find an economically acceptable alternative to the report. This was an impossible task, unless some foolhardy politician was willing to attempt to gut the Endangered Species Act. In September 1990, as the Yeutter task force sought to wring water from a stone, the U.S. Ninth Circuit Court ruled that Section 318's restrictions on judicial review of old-growth timber sales were unconstitutional.[50] This eliminated the possibility of using a legislative rider to set national policy on the northern spotted owl.

Having discovered no way to circumvent the Endangered Species Act or the Thomas report, the Yeutter task force's final report contained little of substance. Fully alert to post–Earth Day citizen sentiment, Congress howled it down. There would be no administration solution to the northern spotted owl problem in 1990.

No solution, of course, translated into no protection. And no protection, now that the Yeutter task force's work was complete, meant that the NFMA and the Endangered Species Act were still being violated. The Forest Service stated that it would manage the Northwest forests in a manner "not inconsistent with" the Thomas report. Environmentalists challenged this quasilegal decision, and once again Judge Dwyer was petitioned for a ruling under the NFMA's viable-population requirements.

In May 1991 Judge Dwyer ruled against the Forest Service on both substantive and procedural grounds. His decision was remarkable for its dramatic clarity:

> The records of this case . . . show a remarkable series of violations of environmental laws. . . . When directed by Congress to have a [decision] in place by September 1990, the Forest Service did not even attempt to comply. . . . This announcement [to manage timber in a manner "not inconsistent with" the Thomas report] was made without notice, hearing, environmental impact statement, or other rule-making procedures. . . .
>
> Had the Forest Service done what Congress directed it to do . . . , adopt a lawful plan by last fall, . . . this case would have ended some time ago.
>
> More is involved here than a simple failure by an agency to comply with its governing statute. The most recent violation of NFMA exemplifies a deliberate and systematic refusal by the Forest Service . . . to comply with the laws protecting wildlife. This is not the doing of the scientists, foresters, rangers, and others at the working levels. . . . It reflects decisions made by higher authorities in the executive branch of government.[51]

The judge gave the Forest Service until March 1992 to develop a plan "to insure the northern spotted owl's viability, together with an EIS," under the NFMA. In the interim, no new timber sales would come from owl habitat in Region 6. On the final day before the court-ordered deadline, the Forest Service announced that it would comply with Judge

Dwyer's request by adopting the Thomas report. The agency and the administration were clearly doing as little as possible to protect the bird.

Dwyer's ruling came on the heels of another successful lawsuit in support of the northern spotted owl. In February, Judge Thomas Zilly had ruled that the Fish and Wildlife Service had violated the Endangered Species Act when it failed to designate critical habitat for the bird upon listing. Two months later, the agency identified 11.6 million acres as critical for the owl—40 percent more habitat than the Thomas report HCAs covered. Did this mean that the government was admitting the inadequacy of the same plan that the administration, the Northwest delegation, and the timber industry had been loathe to accept? In August, the Fish and Wildlife Service offered a revised draft proposal for critical habitat that reduced the area back down to 8.2 million acres. And still the owl had no protection—the proposal would not become final until sometime in 1992.

All of this legal jousting could not obscure three important facts. First, neither the Forest Service nor the Fish and Wildlife Service was going to protect the northern spotted owl unless prodded by the courts. Second, it was difficult to imagine how federal land-management agencies, facing intense anti-owl political pressure, were ever going to succeed in implementing conservation-biology-based strategies with 100-year timelines. The Thomas report was two years old in 1992, and still it had not been put into practice. And third, while the Dwyer decision limited some logging, ancient forests were still falling as the political stalemate deepened.

As if further proof of the biodiversity crisis was needed, another bird and several species of salmon joined the northern spotted owl on center stage. The Fish and Wildlife Service, pressured again by a lawsuit, proposed to list the marbled murrelet as threatened under the Endangered Species Act. And the National Marine Fisheries Service, facing a separate court challenge, listed the Snake River sockeye salmon as endangered after only four fish completed the 900-mile journey inland to spawn. Decisions on other species of Columbia River salmon were pending. Meanwhile, the Mexican spotted owl and the northern goshawk, two indicator species for ancient forests in the Southwest, were both headed for legal listing decisions. These decisions had to be treated separately under the Endangered Species Act, but, taken together, they built a strong case for what ecologically savvy environmentalists had been warning since the release of the Thomas report—a plan to save the northern spotted owl, important as it may be, was not a plan to protect old-growth forests. A strict species-by-species approach to conservation was not going to protect ecosystems. And until the government acknowledged this, one species after another was going to be headed for endangerment.

Stalemate or no, Congress still had the capacity to listen. "This issue began as a dispute over owls and has evolved into a debate over the future of remaining old-growth forests. If we hope to avoid future controversies over new endangered species in this region, we must stop looking at each species in isolation. Instead we must look at the sum of the parts and attempt to craft a strategy which is more broadly based." These were the words of Representative Kika de la Garza (D-Texas), chairman of the House Agricultural Committee, as he met the press upon release of the final report of the Scientific Panel on Late-Successional Forest Ecosystems.[52] Dissatisfied with all the draft Northwest forest bills reviewed in congressional hearings, Representative de la Garza, along with House Merchant Marine and Fisheries Committee chair Walter Jones (D-North Carolina) and two important subcommittee chairmen, Harold Volkmer (D-Missouri) and Gerry Studds (D-Massachusetts), had created the scientific panel as a sort of Thomas committee II. In fact, Jack Ward Thomas was a member of the new panel along with three other scientists: Jerry Franklin, the premier ancient-forest ecologist; Norm Johnson, the father of FORPLAN, a computer model used by the Forest Service in the planning process; and John Gordon, dean of the Yale University School of Forestry and Environmental Studies.

Unlike the Thomas committee, however, this scientific panel went beyond a single species approach to address the design of a regional ancient-forest reserve system. And Congress wanted more. The panel was also to "develop interim protection alternatives for ecologically-significant old growth and late successional *ecosystems*, species, and *processes*"[53] (emphasis added). The panel was to map all old-growth on federal lands within the range of the northern spotted owl and rate it according to how well it protected functional ecosystems, meanwhile developing guidelines for managing unreserved lands. Finally, the panel would evaluate "the effects each alternative will have on timber harvest levels in each area."

It was clear from the beginning what Congress really wanted from the scientific panel—to find out exactly how to protect ancient forest species and ecosystems while cutting the highest possible amount of timber from federal lands. To produce its report, the most detailed regional-ecosystem-scale study ever attempted, de la Garza and his colleagues gave the panel three weeks.

It took Thomas, Franklin, Johnson, and Gordon two months to report back to Congress and several more months to complete their written study. While the ability of the scientists to pull together and analyze a tremendous amount of data in so little time was amazing, the results were predictable. Current forest plans mandating a cut level of 4.4 bbf overestimated the amount of timber available by up to 25 percent.[54] They would do little to maintain the population viability of the northern spotted owl. The panel also found that the forest plans would not ensure functional old-growth ecosystems or provide adequate habitat for the marbled murrelet and twenty-nine other old-growth-dependent species. The ninety stocks (genetically distinct subpopulations) of Pacific salmon analyzed by the panel would also slip toward extinction under the plans.

The Thomas report, the panel found, would, as advertised, provide a high degree of security for the northern spotted owl. But the strategy would do little to protect ecosystem functioning and the other old-growth species examined by the panel. The MIS concept appeared to be a theory disproven; the owl was likely not capable of sheltering other species under its wings.

What levels of logging would still sustain ancient-forest species and ecosystems? The panel provided a range of fourteen basic alternatives for Congress to ponder, but Jack Ward Thomas made the choices clear: "There is no free lunch. Increasing environmental protection comes at a real cost."[55] To achieve only a medium-high probability of protecting all the old-growth elements that the panel studied would lower the cut by roughly 60 percent to 1.3 bbf. The panel estimated that 54,000 timber-related jobs could be lost by doing this. (This figure is somewhat misleading, since implementing the forest plans would alone mean a loss of 26,000 jobs.) A high probability of species and ecosystem protection (likely the only legally defensible choice under the NFMA and Endangered Species Act) would probably send the cut level under 1 bbf, far below the 1990 sales level of 3.7 bbf. If Congress was not willing to achieve this, then the only alternative seemed to be wholesale revision of the most powerful environmental laws in the United States.

Science having done what it could, the members of the panel concluded that "the process of democracy must go forward from here."[56] It would be up to Congress to decide which set of alternatives and which combination of risk levels to select. But although the panelists had provided an invaluable scientific service, they neglected an important detail—they did not clearly delineate the levels of protection below which science would stamp "biologically unacceptable" on the results. Without this, Congress had *too* many options. It was unclear whether legislators, not being biologists, would be able to accept short-term economic loss to prevent the much greater loss entailed in ongoing ecological degradation.

The Genesis of Ecosystem Management

Sometime in the 1960s, while the northern spotted owl's old-growth habitat was becoming fragmented forest, the agencies built bureaucracies, the timber industry mutated into a multinational octopus, and preservationists became environmentalists. Each of these groups saw a different spotted owl. The Park Service, more than others, stood firm. After all, the owl was already protected in the parks. The Forest Service rowed hard against the currents of change set in motion by the new scientific awareness of birds, bears, and ecosystems. The timber corporations saw doom in a small predatory bird, though they might have profited more by looking in a mirror. The image of the northern spotted owl was beginning to teach environmentalists and some politicians to see ecosystems instead of endangered species and biodiversity in place of parks and forests. There were green buds sprouting across the landscape of management, but they had yet to blossom.

As the owl issue developed, some federal land managers continued to offer a false branch of hope about the middle ground, a compromise that would "help society make a more informed and economically acceptable decision."[57] This was a nineteenth-century echo holding out the promise of the greatest good for the greatest number. Other agency planners, understanding the implications of the biodiversity crisis, began to search for new answers. They did not have to look very far.

For the first two days of a week-long interagency ecosystem management workshop near Mount Rainier National Park in the central Cascades, scientists and managers from the Park Service, the Forest Service, and academia debated one of the basic concepts encountered in Biology 101: What is an ecosystem? It did not seem to help that twenty-four of the thirty-three participants held Ph.Ds. *Ecosystem* means different things to different people, now that the concept has become politicized. To Park Service managers of the old preservationist school, it meant the balance of nature. Forest Service managers, cut from Pinchotian cloth, heard in ecosystem a euphemism for "Park Service land grab"—preserving amenities at the expense of multiple-use commodity production. A minority even felt that ecosystems simply did not exist outside the national parks. Scientists, too, were split over the word. To some, ecosystem meant the processes of nature unencumbered by humans, nature as laboratory. Others saw biodiversity as framing ecosystems according to a dynamic pattern-and-process model of life on earth. The implications of the more lucid scientific image of a diverse natural world changing across time and space were revolutionary for land managers historically wedded to resourcism.

Arguing over definitions, the workshop participants could not see that they were wrestling with not just one image of nature but two. Today's evolving scientific view inevitably challenges anthropocentric views of the Earth. There is no place in the world for northern spotted owls or functional ancient forests unless humans give ground.

That first tumultuous workshop spawned several versions of a new landscape of management. Each retains the flotsam of conservation and preservation, bits and pieces of old views of nature caught in the currents of change. Biodiversity protection now competes with commodity and amenity production. As we struggle toward a new image of nature we must ask ourselves, What precisely is ecosystem management?

An old answer to this question is found in the concept of the biosphere reserve. A child of the United Nations Educational, Scientific, and Cultural Organization's Man and the Biosphere (MAB) Program, biosphere reserves were bred in the international conservation think tanks of the early 1970s.[58] There was little that was new in the idea—biosphere reserves would accent protection, scientific research, and monitoring. But the concept generated a provocative spatial model of management: There would be a large protected core (that is, a national park) surrounded by a buffer zone, a restoration zone, and a stable cultural area where "people live in harmony with the environment." Human use would be allowed in the concentric rings surrounding the core, increasing with distance away from the core. This early systems approach did not recognize the dynamic aspect of nature, nor did it treat people as anything more than abstract actors in an idealized world. It was vintage preservationism with a nod at scientific research.

Though today there are over 266 biosphere reserves in seventy nations, it did not take long for the concept to run aground. Most such reserves were simply laid over existing national parks, cores without the intended buffer zones. Conflicts were avoided, but so was integrated land management. After almost twenty years few biosphere reserves fit the idealized pattern, and there are none with fully integrated human communities. Though conceived as a global network to protect examples of worldwide biodiversity, only 1.6 percent of global ecosystem types are represented.[59] Some biologists point out that "within MAB, no thought has been given to studying [ecosystem] linkages."[60] Researchers have not tackled the sticky issue of how big the core area must be to support sustainable levels of human use in the buffers. In the United States, the Park Service has informally adopted the concept as a guidepost to regional land planning. But neighboring national forests have been reluctant to participate, Congress has provided little funding for the program, and the public is hardly even aware that biosphere reserves exist. Overall, this first model of management beyond administrative boundaries suffers from nebulous goals that offer something for everybody. The hard questions are left unanswered. Biosphere reserves may yet have a future, but they are not the panacea that some wish them to be.

Jim Agee and Darryll Johnson published their version of ecosystem management in 1988. Written in scientific jargon, *Ecosystem Management for Parks and Wilderness* is not an easy read.[61] It does, however, contain a fascinating blend of the new and old, replacing the old model of nature as static (balanced) with the view of nature as dynamic. If nature is dynamic, then managers must be flexible. People belong in the equation; managers can no longer discount the effect of humans on ecosystems. "Solutions may not occur if political boundaries are used as ecosystem boundaries," the authors stress; "the scale of the landscape is larger than any traditionally dealt with in natural resources management." In this expanded world, managers must work from a set of clearly stated goals, learn to cooperate across agency boundaries, monitor their work, and be guided by leadership at the national policy level. Agee and Johnson want to "preserve options for the future," and this means protecting biodiversity across landscapes. Their brand of ecosystem management depends upon results, not the number of meetings between agencies or of pages in a land-use plan.

At the same time, Agee and Johnson wrestle with how to pin down ecosystem management in a natural world that is never static. Since what is natural "cannot be scientifically resolved," management goals must rest on achieving "socially desirable conditions." Biodiversity protection is both a primary goal and just one of many social choices. But is protection of the grizzly and the ancient forest a social choice or a biological imperative? Do humans need old-growth forests, do we simply want them, or do these questions merely push us back toward resource conservation?

Soon after *Ecosystem Management for Parks and Wilderness* appeared, biologists Tim Clark and Ann Harvey published a gray booklet entitled *Management of the Greater Yellowstone Ecosystem: An Annotated Bibliography*. It purports to be a review of the scientific literature on management in Greater Yellowstone, but the introduction presents another view of ecosystem management.[62] Tim Clark made his reputation as a biologist studying Wyoming's endangered black-footed ferret. He has also taught wildlife policy at Yale University. He knows ecology, both in theory and in practice, and he is well aware of the role politics plays in both. In an earlier paper, Clark defined ecosystem management as "management of natural resources using systems-wide concepts to ensure that all plants and animals in ecosystems are maintained at viable levels in native habitats and basic ecosystem processes are perpetuated indefinitely."[63] The new booklet builds upon this foundation.

Clark and Harvey share Agee and Johnson's emphasis on data collection and monitoring. They are also forthright about the difficulty of establishing precise ecological boundaries and, like Agee and Johnson, believe that humans have a role to play in demarcation. Social and scientific boundaries will never match perfectly.

The controversy surrounding Clark and Harvey's ecosystem view does not concern the conflict between scientific and social images of nature. What strikes sparks is their statement that all species, not just humans, must be the beneficiaries of ecosystem management. Clark and Harvey challenge wise-use by daring to set standards so that human "use can be sustained without damaging the ecosystem's integrity." This neatly frames the current stage of the biodiversity crisis: Managers are being pressed to push commodities and amenities into the background while protecting the ecosystems from which these goods and services flow. According to Clark and Harvey, conservation biology defies business-as-usual land management. Yet the federal agencies that have grown up to implement environmental policy are not so easily pushed.

Other attempts to define applied ecosystem management pay little heed to the concept of an expanded partnership with nature. In seeking a model of managing for biodiversity, Rich Baker and Christine Schonewald-Cox, working out of the Cooperative Park Studies unit at the University of California, Davis, choose among four strategies.[64] Should we expand parks and reserves? Foster interagency cooperation? Manipulate habitat to increase mammal populations? Or invest in captive breeding programs? Baker and Schonewald-Cox believe that cooperation holds the most promise. It can be mandated legally (as it is through the NEPA and NFMA) or it can be unstructured, bound by administrative agreements instead of law. In most cases, agreements should be as flexible as possible and generated by the agencies.

Hal Salwasser, former head of the Forest Service's New Perspectives program, shares this view.[65] Salwasser believes in a dynamic nature, biodiversity protection, and the necessity of setting clear goals. Unfortunately, the examples of "successful" cooperation these researchers cite do not inspire confidence: the Interagency Grizzly Bear Committee, the Endangered Species Act, and the biosphere reserves. The grizzly is hardly better off after almost twenty years of interagency management. It is taking legal invocation of the Endangered Species Act to goad the Forest Service into protecting the northern spotted owl. The biosphere reserves remain an untested dream. Instead of probing the future of ecosystem management, these researchers have merely outlined the status quo.

What Salwasser, Baker and Schonewald-Cox, and Agee and Johnson have proposed is ecosystem management in a vacuum. Though the insights of conservation biology propel their ideas, and despite warnings that progress must be measured in goals achieved, not the number of meetings convened, theirs is a "process" ecosystem management. Robert Keiter, professor of law at the University of Wyoming, sums up this approach: "While NEPA insures 'process' coordination among neighboring federal land management agencies, it does not insure meaningful substantive coordination sensitive to transboundary ecological realities."[66] In practice, this means few decisions are made outside of the agencies. There is little grass-roots public involvement—concerned citizens are discounted in favor of professional land managers. Environmental laws are seen as burdensome compared with administrative agreements. Process ecosystem management denies history: the Western split between people and nature, the conflict between Pinchot and Muir, the feuds between the Forest Service and Park Service, and the spreading ineffectiveness of political boundaries. As for ethical relations between humans and ecosystems, they are reduced to "socially acceptable conditions," or never mentioned at all.

Agency ecosystem advocates are unable to escape the pull of the federal bureaucracies for which they work. The Forest and Park Services inherited from Progressivism the fundamental belief that big government was the only answer to the excesses of exploitation. Pinchot slowed the destruction of U.S. forests and the Park Service preserved the nation's "crown jewels." One result was the creation of the largest land management bureaucracy in the world. Whether the choice is where to log, how to build a road, who is to be hired, fired, or promoted, or when it is time to hold a hearing or issue a press release, resource conservation created a top-down hierarchy through which filters almost every management decision. "Know your bureaucracy," is the catchphrase of students of the agencies and a key to further exploring ecosystem management.

Ecosystem Management in Practice: Lessons From Yellowstone and the North Cascades

Say a timber sale has just been proposed against the boundary of a wilderness area in your ranger district. Your local forest-watchdog group meets with the district ranger, but she cannot stop the sale. Angry, you write a letter to the chief of the Forest Service in Washington. You demand a review and suggest buffer zone protection for the wilderness and an end to timber-first policies. Months later, you receive a bland reply.

Where does your letter actually go? Who reads it? And who responds? The chief never does. No formal communication channels exist to deliver the general public's concerns to upper-echelon Forest Service decision makers.[67]

Now, say you are the executive vice president of a regional timber-industry consortium. The national forests from which you procure a large percentage of your timber are attempting to lower the allowable cut in response to environmental standards in the new forest plans. You write a letter of complaint to the Forest Service chief. "Jim, this issue was discussed in detail with our regional foresters," replies the assistant chief in a personal letter.[68] He assuages your concerns and promises that the forest plans will not stand in the way of the cut.

Academics disagree over exactly what locks the Forest Service into its timber-oriented brand of land management. Is it the founding mission of the agency, the ghost of Pinchot and Progressivism?[69] Does the Forest Service seek to preserve and perpetuate its budget and bureaucratic organization?[70] Or, as the above example suggests, has the agency been "captured" by its main client, the timber industry?[71] In fact, all of these influences work together to obstruct the birth of ecosystem management.

There is little disagreement over the summary traits of bureaucracies born of the "gospel of efficiency." The Forest and Park Services share a belief in the duty of government to control nature for commodities (Forest Service) and amenities (Park Service), in technocratic (centralized) authority, efficiency and specialization, professional elitism, and maximizing managerial discretion.[72] Work with either agency for long and you soon realize that this is not just an academic laundry list. Advocate biodiversity protection and your lessons begin. What you learn is this: The agencies derive their power from values that separate biodiversity from goods and services, ecology from economics. Management decisions are made in Washington, D.C., and much of the budget is spent there as well. (One district ranger estimates that 50 percent of his resources are spent to "satisfy the needs and whims of . . . the bureaucracy.")[73] Promotions are based on getting the cut out or bringing more visitors to the parks. Information flow is controlled and scientific jargon replaces plain talk in planning documents. In decision making, the agencies are reluctant to relinquish any control.

How can cooperative ecosystem management flourish in such a restricted climate? The answer is simple: It cannot. Process ecosystem management is based on maintaining the balance of power between the Forest Service and the Park Service. Salwasser and the other ecosystem advocates mistakenly believe that cooperation will help each agency to attain its own objectives. In fact, the biodiversity crisis demands that the agencies no longer think in such narrow terms.

It is difficult to deny the weight of history and the built-in inertia of the Forest and Park Services when advocating improved interagency relations. Nonetheless, as the first step along a difficult path, the agencies must learn how to work together if biodiversity is to flourish. There are indications that they are moving slowly in this direction.

Like the North Cascades, the Greater Yellowstone ecosystem is a huge, sprawling land of mountains and rivers without end. Bison, elk, and grizzlies still migrate across thousands of square miles of a rich mosaic of lodgepole forest, wide-open meadowland, and wind-swept alpine ridges. The diversity of the Greater Yellowstone landscape is matched by jurisdictional fragmentation that includes three states, seven national forests, two national parks, and several Indian reservations and national wildlife refuges. At least thirty different administrative players are scattered across 10 million acres of uncivilized country.

Until the late 1980s, there was no comprehensive plan for the area at all. But, as national forest plans for the Yellowstone region were released, it became clear that logging, road building, and oil and gas development would reach new highs. Public concern over management of the nation's most hallowed park and its surrounding forest led to a full-blown congressional investigation and report, published in 1987.[74] The report was critical: There were interagency disputes, grizzly bear data gaps, and too many roads in critical wildlife habitat. Forest plans paid little heed to park plans. The study concluded that existing arrangements were "inadequate for providing complete, coordinated management of the Yellowstone ecosystem."[75]

To ward off new legislation by Congress, the agencies went to work and revived the long-dormant Greater Yellowstone Coordinating Committee. They did not stray far from process ecosystem management. As an initial response to the congressional report, the Park and Forest Services signed an administrative agreement that required "coordinated ecosystem management protection in Greater Yellowstone." The committee then produced an ecosystemwide inventory of resources, hired a regional team leader to develop long range goals, plans, and strategies, and sought public comment on a "vision for the future." However, the agencies avoided using the phrase "ecosystem management" in public and stressed that the administrative agreement should not be looked upon as regional planning. Furthermore, the team leader had no policy-making power and the Greater Yellowstone Coordinating Committee was not legally required to heed the public's advice.[76] Up to this point, such cooperation cost the agencies little. And in the five years following the congressional report, logging and road building increased, grizzly habitat was lost, and the agencies remained in the driver's seat.

By 1988, the coordinating committee had begun to work on an interagency document that would set future management policy for Greater Yellowstone. The committee was co-chaired by Lorraine Mintzmeyer, Rocky Mountain regional director for the Park Service, and Gary Cargill, Rocky Mountain regional forester for the Forest Service. Mintzmeyer, the highest-ranking woman in the Park Service, would later describe the Yellowstone plan, tentatively titled "Vision for the Future," "not simply . . . a regional plan or decision. . . . It was to be a model for interagency cooperation in this area and a model for other areas, well into the next century."[77]

The draft plan was released for public comment in August 1990.[78] It made a genuine attempt to coordinate the goals of the agencies by basing management on naturalness and ecosystem integrity, biological and economic sustainability, and substantive interagency cooperation. It offered a vision of process ecosystem management for the future, and, though it seemed to bring a Park Service–style management to the national forests—there wasn't much chance that the Forest Service was going to change land management in Yellowstone or Grand Teton parks—both agencies stood behind it. Following the management theorists, the agencies apparently believed that cooperation would, by itself, give birth to an ecosystem-scale practice.

Few observers anticipated what happened next. A firestorm of controversy exploded at the public hearings on the draft. Agency staff were called communists and Nazis by private landowners who saw "federal takeover" in the public lands document. The misrepresentations of industry groups had influenced them to speak out. The Wyoming Multiple-Use Coalition, for example, demanded that all mention of biodiversity be deleted from the plan because, "left totally unchecked, the ravages of nature would leave far more environmental destruction than management."[79] The commodity-oriented congressional delegations from Wyoming, Montana, and Idaho all criticized the plan. Representative Ron Marlenee's (R-Montana) statement was typical: The document "essentially writes a blank check for a few wilderness advocates within federal agencies and radical environmental groups to impose their narrow ideological agenda on 17 million acres of land."[80]

The rhetoric and scare tactics of the multiple-use interests rendered civil discussion of process ecosystem management or any other aspect of the plan pointless. At one packed hearing in Montana, when an environmentalist asked attendees to raise their hands if they had read the document, virtually no one did.[81] The debacle over the plan did not end with public hearings. When the final version was released in September 1991, the word *vision* was conspicuously absent from the title. The document was pared down from seventy to eleven pages. Gone was all mention of biodiversity. But the new plan did have clear goals: "This document reinforces the separate missions of the Forest Service and Park Service."[82]

Even as the agencies retreated from their plan, political powers were moving against regional director Mintzmeyer. For continuing to support the original plan, she was summarily transferred less than a month after the final version came out. This action brought Mintzmeyer to testify on political interference before the House Civil Service Subcommittee in Washington, D.C.

Mintzmeyer's testimony was compelling.[83] She stated that her superiors in the Department of the Interior had caved in to White House pressure against the document, and that chief of staff John Sununu had labeled the plan a "political disaster." Since June 1991 she had been in charge of the plan in name only, as Scott Sewall, then a deputy assistant interior secretary, "had 'been delegated by the department' to retain the appearance that the document was the product of professional and scientific efforts by the agencies involved, but that [in reality] the document would be reviewed based on . . . political concerns." Mintzmeyer was careful not to dispute the authority of the Bush administration to control the final version, but, she concluded, "suggesting that the vision document as it presently stands is the result of efforts by the [agencies], based on scientific considerations . . . is, in my opinion, not accurate. To have created this appearance by neutralizing . . . the people who could have continued to guide the document into its being an expression of professional opinions . . . seems to undercut the mandate that the people have given these agencies."

No other U.S. national park carries the clout of Yellowstone. The Greater North Cascades tugs but little at the nation's heartstrings: There has been no congressional study, no Greater North Cascades coordinating committee, and as yet, no agency officials losing their jobs. But the North Cascades do harbor the country's most endangered population of grizzly bears—and more uncut old-growth Douglas-fir forest than any other place in the Pacific Northwest. The stage is either set for a new chapter in land management, or a repeat of old conflicts and evasions.

Process ecosystem management is up and coming in the Greater North Cascades ecosystem. So far, both the Forest Service and the Park Service seem to be adopting it in parallel. The Forest Service's version, as one might expect, is driven by the NFMA planning process: "The Forest Service . . . shall develop, maintain, and . . . revise land and resource management plans . . . coordinated with . . . state and local governments and other Federal agencies."[84] In the Mount Baker–Snoqualmie forest plan, interagency cooperation is identified as a preliminary management concern, then is never mentioned again.[85] The neighboring plans of other agencies are reviewed for possible conflict. Only three conflicts are identified from a total of thirty-five plans. Elsewhere, logging on the border of Mount Rainier National Park in old-growth that the Park Service would like protected is "resolved" by inviting the agency to Forest Service planning sessions. Law professor Robert Keiter believes that the Forest Service reads "consult" where the NFMA says "coordinate."[86] Consultation means that you inform your neighbors of your plans to build roads and cut timber and then you go ahead and build roads and cut timber. Due notice is given, commodities are extracted, and managers maintain their discretionary authority. This is process ecosystem management. And if an agency employee attempts to do more . . . ?

The Park Service version of ecosystem management is somewhat different. There is no NFMA hanging over the head of the agency and the parks have a long tradition of preservation. Greater ecosystems are less threatening to the Park Service. Nevertheless, the transition toward an ecosystem view is not proceeding any more smoothly in the Northwest than it is in Yellowstone. Each park draws up its own general management plan, which outlines policy for a five- to twenty-year period. (The plans are actually written by regional service-center staff in consultation with each park—another bureaucratic procedure.) North Cascades National Park published a twenty-year plan in 1988.[87] Though the park is almost completely surrounded by three national forests, the agencies met for only half a day to discuss transboundary issues.[88] The plan describes the park as part of a greater ecosystem and calls for regional management. And that is the end of it. There are no specific strings attached—no recognition that the Forest Service plans to increase logging on park boundaries, no description of the effects of habitat fragmentation on park wildlife, no discussion of how much overall landscape change is acceptable, no forum provided for the resolution of these issues. The Park Service trumpets greater ecosystems, but it is reluctant to get involved in the sticky business of asserting park protection when threats come from outside park boundaries. This pattern has been found nationwide, in the northern Continental Divide ecosystem around Glacier National Park, the Greater Yellowstone ecosystem, and the southern Appalachian Highlands ecosystem around the Great Smokies, as well as in the Greater North Cascades.[89] The first step in process ecosystem management is to wield the E-word: *ecosystem.* The second step, whether it be establishing a Greater North Cascades coordinating committee or challenging the Forest Service over "derogation of park values," has never been taken in the North Cascades. The lessons from Greater Yellowstone prove that the most difficult steps remain.

Federal Agencies and Public Participation

The Cibola National Forest in central New Mexico is far from the national spotlight. With no neighboring national park, there are few interagency conflicts. This is a land of high desert grassland, sagebrush, and the wide-open skies of the southern Rockies. Only a few well-watered canyons support more than a sparse growth of timber. What the Cibola has in common with the Greater North Cascades and Greater Yellowstone is people who love the land and are willing to fight for its protection. This small national forest shares something else with the Mount Baker–Snoqualmie and Bridger-Teton national forests: a forest plan that emphasizes commodity extraction.

Released in 1985, the Cibola forest plan was one of the first to be completed. The original plan outlined a 300 percent increase in logging levels, ski resort construction on Mount Taylor, a sacred mountain of the Pueblo and Navajo peoples, inadequate watershed protection in a land where water is as precious as blood, and more.[90] Concerned citizens formed a coalition and appealed the plan. After six months of hard negotiating, all on volunteer time, the coalition agreed to drop the appeal in hopes that a newly established Forest Service citizen working group could resolve ongoing differences.

The working group was not successful. After three years the Forest Service has ignored key recommendations of the advisory group. A leading activist relates her frustration: "I'm sitting at the bargaining table with the Forest Service, screaming for the tenth time, 'You don't really think multiple-use means building roads and cutting trees, do you?' And they keep responding, 'Yes, we do.' "[91]

If the agencies are reluctant to work together, they are even less interested in cooperating with the public. Technocracies of professional experts do not mix well with participatory citizen movements. Process ecosystem management discounts public participation—consult any model and you will find no discussion of citizen involvement. This runs counter to every major U.S. environmental law passed since the NEPA. Congress has consistently supported the EIS process, the public's "right to know," and freedom to obtain information. Land-management bureaucracies, however, feel bound only to accept public input, not to use it, just as they will consult with each other but not cooperate. Public opinion is collected as so much data and incorporated into forest and park plan appendices, where it never seems to affect the final decision. Furthermore, the plans are written in such impenetrable jargon that only the most committed "stakeholders" and "interest groups" will wade through it. Randal O'Toole, a forestry consultant who has reviewed more forest plans than any other person, says that "the fundamental assumption behind the planning process is that the main barrier to achieving maximum net benefits from forest management is a lack of information. By collating enough information, including responses from public involvement, planners can help the decision-maker select a plan that is in the public interest. This assumption is wrong."[92]

This illusion of cooperation with the public is not convivial, it is thoroughly economic. It substitutes for human concern for a forest, whatever preferences are identified in market or social opinion polls. The catch for citizens is that even opinion polls don't count. From the northern spotted owl regional guide to the Mount Baker–Snoqualmie forest plan, from New Mexico to Washington, D.C., most people consider the national forests to be treasures. They don't want them cut and they're angry at "the Forest Service [for] letting 'industry' strip them bare."[93] The agencies attempt to manage public participation in the same manner as they manage the land itself. They also assume that the public consists only of Homo sapiens.

Tom Gilbert, a retired MAB consultant now living near the Smoky Mountains in the southern Appalachian Highlands ecosystem, provided glimpses of what may be the most forward-looking cooperative approach at the ecosystem management workshop. Gilbert the scientist, like Agee and Johnson and the rest of the researchers pushing the ecosystem view, takes his cues from conservation biology and its forecast of impending doom for the world's largest mammals. But no one else connects cooperation with the need for grass-roots participation, environmental education, and executive-level leadership. Gilbert's idea of cooperation grows out of a sense of the southern Appalachians as "common ground": People live there, care for the mountains, and "feel the region is their home." He harbors no illusions that progress can be made overnight or that politicians will not attempt to thwart it along the way. But, by calling for interagency decisions ultimately to support "the welfare of all," Gilbert edges managers toward an ethical landscape of management. Just who does "all" refer to—humans? spotted owls? entire ecosystems?[94]

"Who owns the grizzly bear in the North Cascades?" I glanced around a wood-paneled roomful of quizzical faces while gusting rain drummed against the windows and maple branches twisted into knots outside. It was my turn to speak at the North Cascades Transboundary Conference. "Which of us controls the future of the bear? Let me try to answer my own question," I offered. "Maybe we can find some common ground here to help us work through transboundary issues.

"I'll start with my own group—academics. We may not always support the grizzly but we like to think we understand the bear. And we have models and data to prove it. We think that if you managers used our data, the grizzly would be safe. But we forget that our models can't control the bear simply because we believe them to be value-free. So the grizzly slips through. We can't control bears because, without valuing them for *what* they are, we don't really know *who* they are.

"The timber industry, does it control the grizzly? Without the habitat fragmentation caused by logging, the bear might not be in danger. But industry can't support the bear because they concentrate too much on economics—the bottom line. And they don't understand that in ecological decision making, while the bottom line is always important, it's never the primary factor.

"How about the agencies? Let's look at the Park Service. Jon Jarvis tells me that he has no idea how many grizzlies live in North Cascades National Park. How can you manage something that you can't even count? The Forest Service is in the same boat, though from reading the Mount Baker–Snoqualmie forest plan you would think the grizzly is doing just fine.

"What am I trying to say? Grizzly bears are the perfect transboundary animal. Nobody in this room knows much about them, though we're all trying to prove otherwise in our plans and policies. The trouble is, there's no such thing as a Park Service bear or a Forest Service elk herd. There are simply animals moving through habitat, now on Park Service land, now on Forest Service land, now on private land. The animals are simply at home.

"We don't need meetings such as this to validate the existence of a Greater North Cascades ecosystem. The facts of bear biology and conservation science tells us that. What all of us *do* need is a perspective to match the place we manage, a vision that goes beyond our political allegiances to parks, forests, Canada here, the United States there. If we can somehow put the needs of grizzlies, the 'nature of natural ecosystems,' on par with our political motivations, we will have taken a giant step toward ecosystem management. I'm talking about building beyond this our initial meeting. What we need is a willingness to participate, to listen, to offer compliments and critiques, to take some risks.

"Since none of us owns grizzlies, how can we help each other to manage a Greater North Cascades that's bigger than any one of us?"

Silence and the rain on the windows answered.

I sat down. A chair leg scraped the floor. Someone suppressed a cough. People looked down at their hands, glanced quickly around the room, or peered out at the gray sky and sodden trees. For a moment every person in the room experienced a sense of cooperating, working together, encountering the frustration of unfulfilled expectations and the fear of change. These emotions were clearly felt by all present, but it was equally apparent that none of us knew yet how to translate our feelings into the stuff of agency plans and policies. Then one of the group spoke a desultory phrase, broke the spell, and the meeting continued without further discussion of the points I had raised. Twenty-five minutes later, after the regional director of the Park Service offered for the second time to throw $50,000 at interagency management problems, the meeting adjourned and the North Cascades Transboundary Conference was history.

The meeting had failed, no more able to spark reform than a scientific recitation on population viability. There was too much history, hierarchy, and turf at stake for the agencies and the interest groups. I fretted over the ancient forest that was falling even as the participants drove off into the rain. If the meeting taught me anything, it was that the biodiversity crisis would not be resolved overnight.

Reforming the Landscape of Management

Maybe I was wrong.

"Right now, we're all struggling to make a transition. Traditionally, national forest lands capable of growing commercial volumes have been managed largely with a focus on individual stands and what they could yield, especially a sustained or increasing yield of timber. Now the focus is shifting to integrated ecosystem management. . . ."[95] Assistant chief of the Forest Service George Leonard was speaking to an Audubon Society meeting about the changes engulfing his agency, changes he could not control.

In the last several years the biodiversity crisis has rocked the Forest Service like no other force in the history of the agency. In 1989, the Association of Forest Service Employees for Environmental Ethics (AFSEEE) sprang to life from within the ranks of Forest Service field staff. The group's founding statement, the vision of a timber-sale planner and wildlife biologist from Region 6, was an ecological manifesto:

> We believe the value system that presently dominates the Agency in terms of how we manage the land is in need of immediate change. . . . We believe that biological diversity and sustainability is [*sic*] more important than "managing" for fiber forests and short-term political expediency. . . . We believe it's time . . . [to] start meeting the SPIRIT and LETTER of NEPA, NFMA, the Endangered Species Act [and our other resource protection statutes]. . . . We believe we should be designing and implementing land management policies which reflect TRUE stewardship and an ecologically sustainable economic base. . . . We believe the current political and organizational structure of the agency requires those of us within the agency at all levels to start speaking out, for a new resource ethic. . . .[96]

In one year, the new group grew to represent over 5 percent of all Forest Service employees.

Forest Service workers and managers have used other forums as well to express concerns about ecosystem management. A recent agency-wide poll revealed a gulf between what the bureaucracy and what employees believe are appropriate rewards and incentives.[97] The Forest Service hierarchy has traditionally rewarded managers for organizational loyalty, meeting timber targets, promoting the image of the agency, following the Forest Service rulebook, and teamwork. Employees feel they should be rewarded for professional competence, caring for healthy ecosystems, considering future generations, innovation, and risk taking.

Middle-level managers have also counseled reform. Banding together by regions, forest supervisors issued a series of public "memos to the chief" at the annual regional supervisors' meeting in November 1989. The supervisors were forthright: Too much timber was being cut, soil was eroding, there was little regard for wildlife, recreation was underfunded, and in general the agency was "out of control."[98] To solve these problems would take more than new legislation or additional money. The supervisors wanted the Washington, D.C., staff cut and the management chain of command fundamentally restructured.

Supervisors issuing statements, workers organizing for environmental ethics—was this a new Forest Service rising out of the ashes of a thousand clearcuts?

The events of 1991 proved that there would be no sweeping changes in the Forest Service. Administration politicians and top-level agency brass were not simply going to give in to reform-minded regional bureaucrats and field personnel.

Even as Lorraine Mintzmeyer was being muscled out of her Park Service job because of her support of the draft Greater Yellowstone *Vision for the Future*, John Mumma, regional forester for the northern region of the Forest Service (including Montana, Idaho, Wyoming, and parts of Washington and the Dakotas), was forced out of his by what some observers labeled the Timber Coup. Mumma had run afoul of powerful commodity interests represented by congressmen from Montana and Idaho because he was not willing to sacrifice forest-plan environmental standards "to get the cut out." Both Mumma and Mintzmeyer were called to testify about the circumstances surrounding their dismissals before the same House subcommittee. Both testified on the same day. Mumma began his testimony: "I'm here today with a . . . heart that's in shock at what's happening in the national forests of this country." In defending his inability to meet timber targets he stood firm: "I know of no instance in which federal law would have allowed us to cut or offer for cutting more timber than we have done."[99]

The members of Congress who took John Mumma's statement were interested in the political pressure he had been subjected to in the line of professional duty. Mumma spoke of that and also of the uneven spread of ecosystem management nationwide. Just as the scientific panel on old growth commissioned by two House subcommittees had shown for the Pacific Northwest, forest plan timber targets in the northern region were proving impossible to meet for similar reasons—inflated timber levels based on erroneous data, problems with the computer model FORPLAN, and sales set not by foresters but by Congress.

Deputy chief of the Forest Service James Overbay, who brought the news of his termination to Mumma, saw it differently. Overbay believed that "frivolous" environmental appeals were to blame: "When we make a decision to sell . . . a timber sale, we need to stick to it and not re-examine it."[100] Forest-plan timber goals were "intended to be sold and . . . represented commitment [to the timber industry] to be met." His strategy to meet this goal? Relax Forest Service guidelines under the NEPA and the NFMA, provide more funding for timber sales on the national forests, and lobby Congress to limit judicial review of agency activities.

In a letter to Mumma's replacement, chief of the Forest Service Dale Robertson made it clear that all forests must meet their forest-plan cut levels until the plans had been revised following due process under the NEPA.[101] Since revisions might take two to three years, this appeared to be a way to buy time for the timber industry.

Mumma's ouster seemed to encourage Chief Robertson and other important Forest Service officials to move aggressively against agency reform. In testimony before the House Agricultural Committee on the scientific panel's report, Overbay repeated his claim that appeals and litigation were to blame for the unmet timber goals.

At another congressional hearing, Chief Robertson accused the panel of producing a "biased report" because it had not been submitted to the Forest Service prior to publication.[102] It seemed to matter little that three of the four panel members had gained a professional reputation through work done for the agency in the past. And at a press conference in Oregon, Overbay, when asked for his view of Judge Dwyer's legal critique of the agency, said, "We didn't violate the law or harm the owl, we merely got caught up on some procedural problems."[103] In March 1992, in support of Overbay's position, the Forest Service announced that it was going to eliminate citizen appeals of timber sales and forest plans.

With an aggressive Forest Service leadership defending traditional agency concerns, a prodevelopment administration, environmentally unsound logging levels, and no prospect of a solution to the ancient-forest dilemma in sight, the concept of ecosystem management was gaining little ground. But while timber hardliners maintained political control, their victory on the ground was illusory. Where were the logs for future sales to come from? Under the Dwyer injunction the Forest Service had shifted much of the cut in Region 6 to eastside Cascade forests. But the dry-side ecosystems needed their own Thomas report and scientific panel. When Congress finally decided to protect westside ancient forests, the cut couldn't simply be shifted to the northern Rockies; the Timber Coup had proved there was little logging left to be done in Idaho and Montana without violating environmental constraints. Nor could the cut come from the Cibola and other southwestern national forests where large timber volumes had never existed. And, even if they lost the right to appeal timber sales and lost access to the courts, environmentalists and agency reformers were hardly going to fade away as the conditions they were fighting so hard against today became worse tomorrow. Compromise seemed impossible. The middle ground had been cut out from under the northern spotted owl, the nation's timber economy, and Congress.

The landscape of land management is changing, and changing rapidly. Widespread concern over ecological health is fomenting a revolution in attitude of the scale seen earlier in the century with Progressivism. Only this time we are by necessity viewing land management in a larger ecological context.

Conservation biology and current political events suggest that there are good reasons to resist the agencies. Only the blind can survey the western Cascades stumplands and miss the ecological (and economic) implications. In ten years, Greater Yellowstone may be growing more oil and gas wells than grizzly bears. The spruce-fir forests of the southern Appalachian highlands are being destroyed by a photochemical haze. Whole ecosystems across the United States are now endangered; the problem is not limited to the Pacific Northwest.

A scientific ecosystem approach may turn out to be just the latest manifestation of a land-management bureaucracy bent on transforming nature into a human artifact. There is, however, a deeper landscape of management to inhabit: "It's partly because we swept across this continent so quickly, exterminating bison and wolves and so much of our big wildlife . . . that we never really knew what was here. We destroyed the great patterns of migration and interaction. We destroyed the communities essential to our vision of what animals need. By the time we got around to reestablishing any, they were nearly all in tiny, isolated fragments, and that's how we went about managing them."[104]

The biodiversity crisis may encourage us to adopt an ecosystems view, but it should also help us to remember what lived here before, what has been lost, and what needs to be recovered and restored. Beyond process ecosystem management hovers an image of nature that is not only dynamic but also whole, representing all the native voices of North America. Yet current land-management practices, based as they are on what Pinchot took to be the definitive standard, people versus natural resources, blind us to this. The etymological root of *resource* means "to rise again." People cannot be separated from nature any more successfully than the northern spotted owl can be torn from old-growth. There will be no rising again for people, plants, or birds if we perpetuate old standards in a new mold.

If current standards are suspect, what are we to base ecosystem management upon? How, specifically, are we to marry our developing knowledge of species' needs and of a dynamic nature with a new sense of ourselves as part of the community of life? There are two problems here. One is scientific. Our knowledge, though substantial, is incomplete. Any model of management generated by scientific discovery will always be subject to revision as we learn more about species' needs and how nature evolves. The other problem is ethical.[105] We are addicted to the old image of ourselves as separate from nature and for centuries have practiced the form of management that results from it. As recent history indicates, it is not proving easy for us to change. But late-twentieth-century environmental deterioration has forced us toward a sharper focus: The biodiversity crisis requires that we not only redesign nature reserves but also reenvision the values with which we approach nature.

The search for new scientific standards of land management is what conservation biology is all about. But scientific ecosystem standards are revolutionary because they demand that we confront the ethical question of how humans fit in with nature. Surveying the landscape of management makes two things clear. First, though our ecological knowledge is incomplete, it is not a futile exercise to set management standards. The following chapter is my attempt to do so. Second, because the biodiversity crisis requires such a challenging stretch of the imagination, it is essential that we remain open to what scientists are telling us about the needs of owls, bears, and ecosystems of which we too are a part. Managers and citizens must become better learners and they must seek out good teachers. Instruction is not difficult to find—the best teachers are ecosystems themselves. And, as Wendell Berry has pointed out, "the teachers are everywhere. What is wanted is a learner. . . . [The teachers] are waiting, as they always have, beyond the edge of the light."[106]

CHAPTER SIX

ECOSYSTEM MANAGEMENT FOR NATIVE DIVERSITY

Be watchful, and strengthen the things which remain, that are ready to die.
Revelations

If you walk long enough and far enough in the Greater North Cascades, you find yourself crossing borders, worlds of varying terrain. Say I travel from the western lowlands up one of the broad, flood-plained river valleys, over the Cascade crest, then down the eastern slope to Columbia River sagebrush. What do I see as I walk? Where am I located ecologically?

For the first sixty miles of my journey I travel in the Skagit River lowlands. Green patchwork fields of corn, hay, and other crops surround small service-center towns. I pass through swatches of second-growth Douglas-fir, western hemlock, red alder, and fencelines full of blackberry brambles and European weeds. The river winds through this landscape, guarded by a narrow riparian corridor of cottonwood, willow, and alder. Asphalt and gravel roads seam this quilt of made-over ecosystems. My botanist's eye picks out the immigrants from the native plants, categorizing genus and species. I recall the taste of spearmint tea and blackberry pie. My feet know the difference between roadside gravel, border tangle, and spongy summer fields of grass. I can see fresh clearcuts high up the sides of the valley spreading through second growth like a mange. Riverbottom old-growth is an oldtimer's memory.

My imagination fast-forwards into the future. Remove people from this valley and in fifty years fields become forests, riparian woods mature, the hillsides begin to heal, and hordes of salmon return to spawn. Corn and alder turn into fireweed and Douglas-fir, a mint patch becomes salmonberry and cedar. Towns fade, their edges invaded by weeds, shrubs, and young saplings. Roads grow grass. Bears sneak out of the hills to fish. On the westside, each ecosystem moves toward some mix of forest, the time to canopy closure based on its prior life as backyard, potato field, or wood lot. Succession leads the patchy pattern until windstorm or wildfire resets the pace. Further in the future, ice or atmospheric carbon augur new changes.

In near time, it is likely that people will remain, towns expand, and logging and farming continue. But my knowledge of ecology and the limits of growth do not permit me to believe that current patterns will remain for long.

Near Rockport, the Skagit picks up steam as the land tilts upward. Here I encounter ancient forest, the only large stand remaining in the valley. It is by chance and not design that this fragment of native diversity survives this far down from the mountains. A 1,000-acre old-growth remnant was willed to the state early in this century and became Rockport State Park. Developed for car camping, most of the patch remains undisturbed in comparison to the rest of the lower Skagit watershed.

This is old-growth that fits the ecological definition: large trees hundreds of years old, a multilayered canopy, plenty of snags and down logs in various stages of decay. My eyes adjust to the dim, green-filtered light. One hundred years ago this forest patch was woven into the larger fabric of ancient forest that stretched across the Skagit and south as far as Oregon and northern California. It climbed the now-clearcut slopes of Sauk Mountain, stitched into forests running north through British Columbia. To the east it was part of a pine marten's hunting path up to the limits of continuous forest and the beginning of the Cascades' subalpine flower gardens. West it extended without interruption to salt water. The regional boundaries of the past are now submerged under generations of human use. Old-growth still exists, but it is severely fragmented below 3,000 feet and scattered in disconnected patches across the western Cascades. Any return to a healthy ancient forest on the landscape scale awaits management decisions yet to be made, followed by at least a hundred years of regrowth.

On the scale of the Skagit Valley, the Rockport forest is an anomaly, an endangered ecosystem fragment. It is too small to support northern spotted owls for much longer, though its roughly square shape dampens the edge effects that would render the interior of the stand less habitable to other old-growth-dependent species. It is completely severed from the nearest ancient forest by many miles of cutover land, a classic island in an anthropogenic sea. But, deep in the forest, if I confine myself to my immediate surroundings, I recognize a ten-foot-wide red cedar, two huge Douglas-firs, and several down logs covered with a rich green tangle of devils club and sword fern. Which world is best to focus on—the stand, the valley, or the Pacific Northwest landscape?

Two days travel eastward and I am bridging the gap between Rockport State Park and the primeval forests of North Cascades National Park. The distance is farther than a juvenile spotted owl can fly. Now I am passing into federal lands. The Skagit narrows to a rocky gorge, the clearcuts disappear, and the steep topography telescopes community boundaries. I walk through an old burn on the south face of a nameless ridge where once lightning struck a fire that burned to bedrock. Silver snags and lodgepole pine say that the soil is rocky and thin.

My roadside path becomes a mountain trail. Douglas-firs are nowhere to be found. I ascend into midelevation Pacific silver fir and mountain hemlock snow forests where winter lies deep and long. These are recreational lands. Everyone I meet is on weekend time, hiking, fishing, and camping. Most visitors do not stray far from their cars, though twenty years back there wasn't even a road up here. Now there are deer roadkills, scenic pullouts, and multicar trailheads.

Turning up Granite Creek I leave people behind, and for the next three days I don't see a soul. For half a day I climb an old miners' track, pass a played-out gold claim, then I'm wrapped in wildness, off trail, heading into the heart of the mountains. Science tells me that the high country has less species diversity than the lowlands. The Forest Service sign I passed in the morning warned that motorized vehicles are forbidden in designated wilderness. I remember from a map at home that this country is large enough for grizzly bears. If I hiked north I could walk for weeks into Canada and never hit a road. Each night, under a blanket of stars, the rest of the universe seems right next door. Morning breaks to damp earth, dewy grass, and, high in the sky, jet contrails tracking the upper atmosphere.

I continue east, drawn by the dry air and scent of pines. All the streams now drop below me toward the rising sun, the other side of the mountains. I'm traveling with Douglas-fir once again, only on the east side they're smaller than their western relatives. Their bark is ash gray and not so ruddy; chartreuse wolf lichens replace the *Lobaria* species of the wet country. Vestiges of innumerable fires show in catface treetrunk scars, soil charcoal, and different-aged forest stands across the U-shaped sweep of the upper Methow River valley. Wildfire perpetuates shifting boundaries; it eats the accumulated dross of the seasons more quickly here than on the damp west side.

At 2,800 feet, I find the first ponderosa pine. It stands alone with no other of its kind in sight, at the upper limits of its tolerance for cold and snow. Maybe a mountain chickadee or a Stellar's jay dropped a seed here forty years ago and that year the snow melted early and the growing season was long and mellow. The next pine I find is a quarter mile away and several hundred feet lower in elevation.

After the pines come sagebrush and snowberry, bittercherry and buckbrush, a logging track, a ruddy weathered cabin, and soon the sound of the highway. Then I'm striding across a flat hay meadow down a long valley between rocky, timber-strewn slopes in bright sunlight. There are groves of aspen hiding deer and long shadows sliding down from the western crest. In four days I will bathe in the Columbia River, sixty miles away. By then I will have eaten flapjacks in three towns full of loggers and tourists, hopped twenty-two barbed-wire fences, been chased by two steers and one angry dog, and walked through federal park and forest, state, and much private land. In all this valley, which once bred grizzlies and 200-foot trees, there is not a shred of ancient forest, no uncut stand where the evening light can glance off a "yellow belly" pine trunk. No place left for the rarest forest ecosystem in the North Cascades, where a person might discover what it means to be old, wise, and wild.

My walk across the Greater North Cascades takes me three weeks, while the same trip by automobile might take a day. The Seattlite behind the wheel, bent on a weekend destination, may not perceive the Skagit Valley as damaged habitat, though she may be angered by the preponderance of clearcuts slicing up surrounding hills. The Rockport old-growth, a thin band of ragged trees boardering the highway, probably remains unknown.

A logger from down valley who spends his work week in the woods sees both a threat and a promise as he guides his rig past the giant timber. The threat comes in the form of the northern spotted owl hovering on silent wings over his job prospects. The promise is that someday the Rockport forest will be grist for the chainsaw and money in the bank.

Three diverse experiences of the North Cascades: an ecologist rambling through space and time, an urbanite escaping the city for a much-needed respite, a worker worrying about where the next paycheck is going to come from. Three ecosystem images, each with its own set of shifting boundaries, cultural and biological. Where do we take a stand? When humans come into conflict with biodiversity, what management standards should prevail? Which image is the "truth"?

There aren't really rights and wrongs in ecosystems, but there are certain biological and ecological truths that must guide the setting of new management standards. Our old standards have never been explicit. We have spoken of wise-use, but our management practice has not reflected that. Multiple-use sustained yield has given us single-use resource depletion—forests cut down, rivers dammed, and grasslands grazed away. Amenity preservation has resulted in parks as ecological islands, crown jewels without a crown. The history of land management should alert us to the fact that we have not accomplished what we set out to do even in the most limited sense—protect natural ecosystems so that future human generations can also be sustained by them.

It is time to untangle the "scientific" search for "correct" boundaries from inherently value-laden questions about management goals. We must recognize that the choices we make not only reflect values but also limit the direction of our hands-on work with nature. Goals and boundaries must be based on the best scientific understanding we have of the limits that come with being one more dependent species in a world of diversity. Ecology has given us relationship as a scientific model for how life on Earth works. We may choose to ignore the metaphorical implications of ecosystem patterns and processes, but we cannot for long escape what science, for now, is telling us about nature. Conservation biology contributes to the field of ecology by exploring the biological bottom line at various scales: extinction of species, viable populations, endangered ecosystems, landscape-level disturbance patterns. The biodiversity hierarchy may not literally exist, may be no more than a "temporary expedient for a particular analysis," just another human image of nature.[1] That is not the point. The point is that ecology and biology are epistemological tools, our best shot, for the moment, of describing, understanding, and fitting in with the world. We know that the risk of extinction increases under definable conditions, that wildfires cannot be long suppressed without significant successional consequences, that certain species play more important roles in ecosystems than others. We are also coming to realize that our values have for so long prevented us from putting our ecological knowledge to work that we are now hard up against the limits of life on Earth for many species. Where we once thought endangered species were the problem, we now face the loss of whole ecosystems. Local and regional environment issues have grown into global concerns.

Some observers posit that ecological degradation results from a "bad way of life." Certainly, there is currently no principled basis for deciding ecosystem management standards, unless by principled one means political and/or economical expediency. Western industrial societies have consumed nature, and the fight for the leftovers has begun. There are few resources left for developing countries to exploit, and precious little habitat for cheetahs, bears, tigers, and tailed frogs. Anthropologist Roy Rappaport distinguishes land-management science from land-management values when he reminds us that these "anthropocentric trends . . . may have ethical implications, but the issue is ultimately not a matter of morality or even of *Realpolitik*. It is one of biological viability."[2]

Ecosystem-Management Goals

In describing new scientific standards of land management I would begin by saying that, at the least, our standard way of life is unecological. We need new norms, and they must be clearly defined and articulated. We may already have them in two definitions offered by Jim Karr and others not long ago. *Biological integrity*: "The capability of supporting and maintaining a balanced, integrated, adaptive community of organisms having a species composition and functional organization comparable to that of the natural habitat of the region."[3] And *ecological health*: "A biological system—whether it is a human system or a stream ecosystem—can be considered healthy when its inherent potential is realized, its condition is (relatively) stable, its capacity for self-repair when perturbed is preserved, and minimal external support for management is needed."[4]

Note that these standards are based on what we know nature produces, not what we desire nature to produce for us. The standards are serviceable at all scales, local, regional, global, because they embrace an ecosystem perspective. Bryan Norton describes this as contextual management: We can see specific problems as well as the context of problems.[5] With ecosystem integrity and health so defined it is difficult to limit our focus to one logging road, endangered species, old-growth fragment, national park, or lightning strike. The big picture becomes the focus.

Given what science is telling us about protecting biodiversity, these general standards reveal four main ecosystem-management goals or conservation-biology criteria:

1) To protect enough habitat for viable populations of all native species in a given region
2) To manage at regional scales large enough to accommodate natural disturbances (fire, wind, climate change, etc.)
3) To plan over a period of centuries so that species and ecosystems may continue to evolve
4) To allow for human use and occupancy at levels that do not result in significant ecological degradation.[6]

This is the scientific heart of solving the biodiversity crisis. These goals, when put into practice, would act as "pre-emptive (ecological) constraints" on the search for human economic welfare.[7] Anthropocentrism and ecosystem science don't mix. The chasm between how we manage now and how we must begin to manage very soon is wide and deep.

Putting ecosystem management into practice requires more than science. Even ecological norms come with ethical strings attached. Those strings become knotted when we attempt to implement management goals. Any outline of a new ecosystem-management model goes beyond ecology to include the federal agencies, Congress, state and private ownership, and scientist and citizen participation in decision making. We must continue to ask this question: Is ecosystem management for native diversity sufficient to sustain biodiversity over time?

History tells us that fundamental change does not come easily, peacefully, or in a planned manner. As we begin to envision more serviceable land management, two problems become obvious. The first arises out of the gap between where we are (multiple-use) and where we must travel (ecosystem management). So much ecological destruction has resulted from resourcism that we have no alternative but to practice hands-on management. There is much damage to repair, many roads to close and forests to replant. Restoration is critical now, though it may not be an important goal later on. Short-term strategies may bear little resemblance to long-run ecosystem management. The danger is, of course, that our current values will subvert short-term hands-on research and restoration and we will end up inhabiting an overmanaged planet, a kind of natural zoo.

The second problem is that, though there are norms for ecosystem management as reflected in the four conservation biology criteria, they cannot be implemented across the globe. There is no planetary method of guiding *specific* management practices any more than there are exact ecosystem boundaries to plot on a map. What works in the North Cascades will not likely succeed in Yellowstone, California, or Kenya. The status quo standards of resource conservation must not be replaced by another narrow standard, even if it appears to provide for more sustainability. As ecosystem management matures, it must derive from national guidelines but defer to regional and local knowledge of the land. This will not happen soon. The following sketch of an interim type of ecosystem management is offered as a bridge-building strategy toward that future. It is only the beginning of what might be a long, rich, and sustained conversation with nature that few citizens of industrial societies have entered upon.

Outlining a Model of Ecosystem Management

How much habitat does a healthy population of grizzly bears need in the North Cascades? What scale of management would provide enough territory for 1,000 gray wolves to roam? No one knows the answers. There is little theory, little data, no magic numbers. Parks and reserves are too small and too few, forests are fragmented, and human populations continue to expand at the expense of native diversity. Given present conditions and trends, how can managers put into practice the new goals of viable populations for all native species and functional natural-disturbance cycles over a planning period of centuries?

Managers must see the endangered species and ecosystems they have inherited with new eyes. Conservation biology theory and research have provided a better understanding of the different levels and scales at which biodiversity must be considered: genetic diversity within populations, the number and kinds of species and populations, ecosystem diversity, as well as ecological patterns and processes that knit these considerations into functional wholes across space and time.

Conservation biologist Reed Noss illustrates management for biodiversity: "To save the Florida panther (or any species one chooses) we must consider the genetics of inbreeding, the impacts of disease and prey scarcity on individual health and population viability, the pattern of patches and corridors across landscapes, and regional habitat changes that may result from climate change."[8] In effect, managers must learn to view species not just as individual units but also as parts of the ecosystems upon which they depend. Furthermore, managers must consider ecosystem interactions at regional and global scales.

In other words, think big and think connected. Bigness and connectivity across our wounded landscapes must be bolstered by a moratorium on habitat degradation and more conservation-biology research. In the present political landscape, bans on old-growth logging coupled with expensive research, expansive wilderness, and biological corridors are considered utopian. That may be so. But if they are, then we had better be very clear about the consequences of pragmatism for both species and ecosystems: They will soon disappear, along with Earth's habitability for Homo sapiens. There are alternatives, if we wish to consider them. Here is a working sketch of ecosystem management as it might be applied in the Greater North Cascades over the next hundred years.

In the Pacific Northwest, from Northern spotted-owl studies to landscape-scale considerations of old-growth, there is not a single scientific justification for the continued cutting of ancient forest. To begin implementing ecosystem management there must be a complete moratorium on all old-growth logging on private, state, and federal lands on both sides of the Greater North Cascades. Roadless areas and unprotected wilderness should also be off-limits to development activities. (At minimum, this should apply to all federal lands.) The time for this ban is now. Yet protecting remaining old-growth and roadless lands is not enough. The loss of over 90 percent of the most widespread regional ecosystem in the Pacific Northwest cannot be corrected by an eleventh-hour decision. We need to grow back more ancient forest by allowing younger stands to reach 200 years of age or more.

How much ancient forest needs to be restored? What is a healthy balance of roadless versus developed lands in the Pacific Northwest? In the United States? Answers to these and many other questions must be based on ecological research that has not yet been initiated. Step two of implementing ecosystem management for native diversity involves funding major research in conservation biology. This new initiative would not replace existing programs such as endangered species and nongame-wildlife research. Instead, it would provide increased monies for current species and ecosystem research and augment these with studies focused specifically on biodiversity at state, regional, and national scales.

Dr. Mike Scott and his associates have found an excellent way to define, catalog, and map hotspots of biodiversity without spending exorbitant amounts of money.[9] Scott's inventory method is called gap analysis. It is based on a simple premise: Species, ecosystems, and landscapes that are not found in parks, reserves, or wilderness areas represent gaps in our protection of biodiversity. We need to find these gaps and fill them.

Gap analysis would be easier to perform if there existed a standard classification of continental ecosystems. But none exists. Or, rather, there is no consensus about which of many classification systems is appropriate. Take California, for example. Depending on where you draw ecological boundaries, whether you are a community-type "lumper" or "splitter," there are anywhere from 15 to 375 natural ecosystems in the Golden State. Biologists typically use vegetation in classifying ecosystems because it is easy to measure and map; plants don't run away from researchers. But establishing boundaries is only one problem in classifying communities. Most such schemes use only dominant, *potential* vegetation as their criterion, based on knowledge of climate, soils, and presettlement conditions. The trouble is that for most of the United States, potential natural vegetation no longer exists. And, just as a single-species approach to biodiversity protection may fail to account for landscape-scale influences like fires and fragmentation, current community classifications treat vegetation types as so many homogeneous units. But ecosystems, unlike many species, directly reflect environmental gradients—moisture level, soil type, amount of sunlight, and so on—which change across the landscape. When gradients are steep, ecosystem boundaries can be very distinct. One hundred feet away from a streamside cottonwood forest, desert grassland may prevail. Where environmental gradients are more gradual, however, defining ecosystem boundaries is somewhat arbitrary.

Scott does not use vegetation alone for classification. To capture unprotected hotspots of biodiversity, gap analysis combines actual vegetation patterns; the location of terrestrial-vertebrate centers of species richness and endemism (where species are restricted in geographic range); the location of the same for terrestrial invertebrates (as represented by butterflies); and the distribution of all federal- and state-listed endangered, threatened, and sensitive species. These categories take advantage of existing knowledge. Land mammals, butterflies, and threatened and endangered species of all kinds are relatively well known. Together, they provide a composite view of where biodiversity protection has been successful and where work still needs to be accomplished. The data is fed into computers using a geographic information system (GIS) program that analyzes multiple layers of geographically based distribution information. A series of maps is then printed showing a composite picture of nodes of diversity in relation to federal, state, and private ownerships, the spatial distribution of biodiversity hotspots, and where corridors could link hotspots across administrative borders. Maps can also be produced that show alternative management and protection strategies. The species and ecosystems most at risk can then be prioritized (though this begs the question of pitting wolves in the Adirondacks against grizzlies in the North Cascades). How much protection would serve *all* elements of biodiversity? What ecosystems would be lost if only 50 percent of the gaps are filled? Eighty percent? Gap analysis can go a long way toward providing the ecological information with which to answer such questions.

Scott and his research team have already applied gap analysis in Idaho. Projects are almost completed in Oregon, California, and Utah, with work in eight more states under way. Gap analysis, however, is not the final answer to biodiversity questions. It is a quick and dirty inventory to collate as much information as possible in a crisis situation. If implemented quickly on a state-by-state basis, gap analysis could target hotspots throughout the United States in five years' time. But analyzing old research results in new ways is only the beginning.

There is also much basic research to be done. The Society for Conservation Biology has published an entire book on research needs alone.[10] There are many species at risk about which little is known. How many bears and northern spotted owls actually live within the Greater North Cascades ecosystem? What are the competitive dynamics between spotted owls and nonnative invading barred owls? Where do most marbled murrelets nest? Are wolverines extinct in Washington? At what road densities are wide-ranging carnivores driven to wilder country? What percent of a watershed can sustain logging without unacceptable erosion? Is group-selection cutting better or worse than clear-cutting? How wide do wildlife corridors need to be for elk? Deer? Bears? Cougars? Where should corridors be located to facilitate the movement of as many animals as possible? What is the minimum dynamic area for eastside lodgepole pine forests? Ponderosa pine? Is North Cascades National Park large enough to encompass a functional minimum dynamic area, or does the park need to be expanded? The list goes on and on.

Forest Service and Park Service researchers may be equipped to answer some of these questions, but there must be increased funding from the National Science Foundation and other large federal and private institutions. Current funding levels are abysmal. As of 1990, only 1 percent of all federal science funding went to environmental biology.[11] The passage of H.R. 1268, the National Biodiversity Bill, would be most welcome. State agencies should also support biodiversity research.

With a moratorium in place, results from gap analysis coming in, and long-term studies under way, the third step in ecosystem management would have a firm scientific footing. Creating a biodiversity-protection network would involve expanding existing reserves as well as restricting human activity in new areas identified through gap analysis as important to the maintenance of ecosystems. The political conflict surrounding the establishment of such a network would make the northern spotted owl controversy seem like a gentle prelude. But, at the least, more sophisticated conservation-biology analysis would paint a picture of what could be lost as well as gained as various proposals were challenged by those who do not make the connection between biodiversity and human welfare.

Biologists Reed Noss and Larry Harris have a biodiversity-protection-network model that would help with this task. The concept is more elegant than its name: the Multiple-use Module (MUM).[12] A refined version of the biosphere-reserve model, each MUM would have

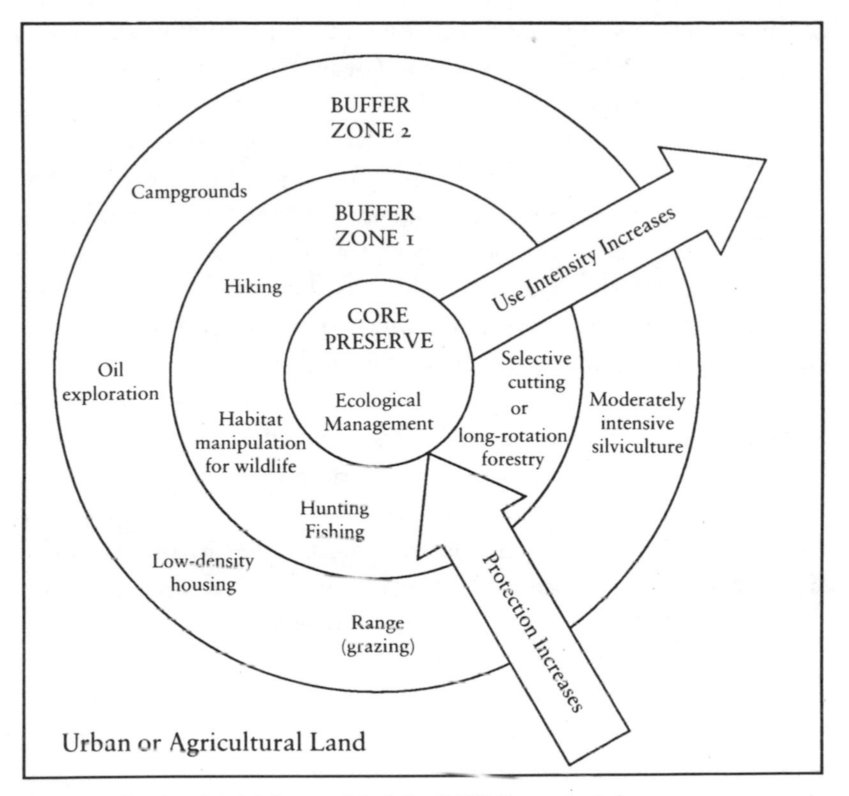

FIGURE 6 In this Multiple-use Module (MUM), an inviolate core preserve is surrounded by a gradation of multiple-use buffer zones. Intensity of use increases outward through the buffer zones, while intensity of protection increases inward. Important functions of a MUM are to (1) insulate sensitive elements in reserves from intensive land use and other human activities, (2) provide marginal habitat for animals inhabiting a reserve, which would increase effective reserve size, and (3) provide for an assortment of human uses with minimal conflict.

Source: R. Noss, "Protecting Natural Areas in Fragmented Landscapes," *Natural Areas Journal* 7 (1987), 6.

at its core a biodiversity hotspot, national park, wilderness, or other protected area (see figure 6). This core would be surrounded by a series of concentric buffer zones with human use increasing outward. MUMs would surround each node of the biodiversity-protection network. They could be designed to protect diversity at all scales, from a local population of calypso orchids or tailed frogs, to the watershed pattern of uncut old-growth in the Skagit River, to representative samples of regional ecosystems throughout the Greater North Cascades. A national network would be built from the system of regional reserves.

One MUM research question looms large: To retain native diversity, how big does the core need to be in proportion to the buffer zones? And the flip side of this question: How intense can human use become in the buffers before the core is no longer protected? The MUM model is incomplete until this balance is addressed quantitatively through research and experience. A tentative working hypothesis might retain at least 50 percent of the entire MUM in protected core.[13]

The next step would be to connect each part of the biodiversity-protection network through corridors to facilitate the movement of species. No MUM is an island. Even the largest reserve or biodiversity hotspot is, in the short term, going to be subject to continuing stress because of existing habitat fragmentation. After the turn of the century, increasing greenhouse gases are likely to cause massive landscape shifts and migrations for numerous plants and animals.[14] Species on the move will need to have somewhere to go, and the corridors connecting the biodiversity-protection network would provide part of the answer.

We know next to nothing about the ecological specifics of designing biological corridors. This should come as no surprise. A society that has, so far, paid little attention to constructing an adequate park and reserve system built around individual units cannot possibly have explored how to link them. But connect we must. Given continuing development and the political costs of enlarging the current reserve system, designating corridors may become one of the *first* steps toward ecosystem management. This is based on the untested assumption that corridors could sustain some level of use and would not require the full protection afforded core areas.

However corridors function, they would need to facilitate movement at many levels. Small mammals and some birds may be assisted by narrow fencerows, but these would not suffice for eagles, bears, wolves, and other large, wide-ranging species. Managers must begin to evaluate all opportunities: unprotected roadless lands, coastal strips, ridgetops, riparian zones, urban greenbelts, power lines, and highway and railroad rights-of-way. What size of biological corridor would provide adequate space for climatically induced continental-scale shifts in species? This is a daunting question. Only time, research, and experience will answer it. Meanwhile, managers, planners, and biologists would be wise to remember the difference between *establishing* corridors and *maintaining* native landscapes and ecosystems that have been functioning as natural corridors ever since the glaciers retreated 10,000 years ago.

Ecosystem managers need to move faster than glaciers. To make a biodiversity-protection network functional, corridors need to be designated within the same five- to ten-year period as MUMs. In the Greater North Cascades, the Thomas committee plan offers a corridor-based scheme for a single threatened species, the northern spotted owl. Other species have different habitat and movement needs, as the Scientific Panel on Old-Growth report has underlined. Faced with so little knowledge about how most North Cascades species move, planners should begin by maintaining landscape connections that remain intact today. Many potential corridors are threatened by proposed logging and road building; the first step toward ecosystem management, the moratorium, would buy precious time to evaluate these areas. Fine-tuning would come later as research results trickled in. The scientific panel suggested three years as an interim research period for ancient forests in the Northwest.

A ban on logging in ancient forests and roadless lands would also give the Forest Service time to learn how to work with forests instead of fiber farms. Ever since Gifford Pinchot imported European silviculture to America, forestry has focused on how to cut down and regenerate trees efficiently, paying little heed to ecosystem structure and function.[15] In the Pacific Northwest, the Forest Service abandoned research on old-growth in 1958.[16] Agency silviculturists had by that time learned how to produce the maximum volume of board feet from what were seen as "decadent, biological deserts." Their method? Clearcut the ancient trees as quickly as possible, burn the leftover slash, replant genetically selected Douglas-fir in plantations, apply herbicides to keep "weed" species like red alder suppressed, and plan to cut the next stand just when the quick growth of youth—in forestry lingo, the "culmination of mean annual increment"—began to level off. Since culmination takes place on most sites in the Northwest after eighty to one hundred years, the forests of the future would never be allowed to become ancient forest again. Paradise for a Cascades forester was a homogeneous monoculture of thrifty Douglas-fir. There was nothing more to learn.

Establishing forestry that sees forests as well as trees will take sustained research and experimentation. Until recently, most old-growth research funding has come from the National Science Foundation, not the Forest Service. Dr. Jerry Franklin, professor of ecosystem analysis at the University of Washington and chief plant ecologist with the Forest Service's Pacific Northwest research station, has been the preeminent student of old-growth for over thirty years. In the early 1980s, Franklin began developing an alternative silviculture that has become known as new forestry.[17] Franklin envisions new forestry at both the stand and landscape levels: Foresters would work with a particular site as well as the landscape context of the site. At the stand level, Franklin suggests retaining green trees, snags, down logs, and other woody debris instead of clearcuts and slash burns. He calls these structural elements of ancient forests "biological legacies." New forestry pays attention to what is left as well as to what is hauled away. At the landscape level clearcuts might be reduced, rearranged, and clustered to reduce fragmentation. Current practices spread cuts across watersheds, maximizing edge habitat and creating the checkerboard effect known to anyone who has flown over the Cascades on a clear day. New forestry also considers the cumulative effects of logging in any given drainage. Before roads are built the interplay of cut and uncut stands would be studied so that biological corridors could be protected. On the regional scale new forestry would bring conservation biology into play by maintaining a "dialogue" between commercial forests and reserved areas. Following the MUM model, Franklin's alternative forestry might someday govern the landscape within which the biodiversity-protection network was embedded.

New forestry is to silviculture what conservation biology theory is to reserve design. Both seek to readjust management radically and bring to bear on it the latest scientific information. Both disciplines are revolutionary, both are meeting great resistance, and both are prone to misinterpretation by those who feel threatened by the implications of scientific ecosystem management. One thing is clear—there can be no biodiversity-protection network without a sustainable landscape of use as well. Ecosystem management is about both working with nature and letting it be; it is hands on as well as hands off. The Forest Service is building new forestry into its New Perspectives management program. At this stage, however, new forestry is still a dream in Jerry Franklin's eye. Time, research money, experimentation, and plenty of political support are needed before it can help to create use landscapes that will also support a functioning biodiversity-protection network.

Another important step toward ecosystem management would be ensuring that ongoing research in new forestry and conservation biology is used to fine-tune initial biodiversity-protection network designations. For all their theories of feedback loops, biologists do not have a good track record of conducting long-term studies that monitor initial research results. Poor monitoring has plagued the Forest and the Park Services. Say, for example, that biologists determine grizzly bears need a southern travel corridor through the Twenty Mile–Thirty Mile roadless area of the Okanogan National Forest, on the east side of the Greater North Cascades. Once the corridor is established, there is no followup to determine how well the design is serving the bear. It has always been difficult to obtain funding to monitor research. This must change. Park Service scientist Gary Davis has already outlined an excellent monitoring program centered on Channel Islands National Park in southern California.[18] This approach could usher ecosystem-management research out of the "balance of nature" dark ages and into the modern world, where nature is viewed as dynamic and forever changing.

The final step in initiating ecosystem management would be the recovery of those species and ecosystems that have already been severely damaged as well as those at risk of slipping away. The end of interim ecosystem management, maybe a hundred years hence, would see the restoration of habitat and ecosystem patterns and processes fit for viable populations of all beings, human and nonhuman alike.

Restorative ecosystem management is rooted in two observations, only one of which has the authority of scientific consensus behind it. Conservation biology has made clear that there is not enough protected habitat for at least most of the largest animals in the world. The ecological consequences of losing many of the planet's whales, bears, big cats, and other top-level carnivores are not completely known, but predictions are grim. Building a biodiversity-protection network will not resolve this problem. Too much habitat has been destroyed or fragmented. The only prudent scientific response to the biodiversity crisis is to restore some balance between what humans have appropriated from nature and what nature needs to maintain functional integrity. Integrity, from a narrow human perspective, might be defined as the continued provision of the "ecological goods and services" that we depend on for survival.[19] Loss of a functional Pacific Northwest ancient forest ecosystem or protecting a mere 5 percent of North America as wilderness is not scientifically justifiable. Long-term ecosystem management means addressing these imbalances by adopting a wilderness recovery strategy: closing and revegetating roads, allowing wildfires free play to set successional rhythms, removing settlements from sensitive areas, reintroducing grizzlies, wolves, and other extirpated species, creating polyculture agroecosystems, and in general building a restoration economy.

Restoration entails puzzling out presettlement patterns of vegetation. There also needs to be some yardstick by which to judge recovery efforts. Historical accounts, early land surveys, pollen evidence from paleoecology, and soil types can all contribute to a composite picture of healthy ecosystems prior to Euro-American settlement.[20] From such data, and taking into account Native American land use where possible, biologists and managers could construct a computer model to track how vegetation changes over time without excessive human interference. Managers could use this as a rough empirical standard of ecosystem integrity to set further management goals.

It is important to distinguish restoration from reclamation, as biologists William Jordan, Robert Peters, and Edith Allen make clear: "By restoration we mean the recreation of entire *communities* of organisms, closely modeled on those occurring naturally, and we . . . use the term *reclamation* . . . to refer to any deliberate attempt to return a damaged ecosystem to . . . productive use or socially acceptable conditions short of restoration."[21]

Recreating pristine presettlement conditions would certainly not be the goal of ecosystem management. No presettlement paradise ever existed. Instead, restoration would be based on understanding how long-term natural climate changes and shorter-term disturbances pattern landscapes. The specific goal would be to allow native species, ecosystems, and landscapes to evolve in response to a dynamic Earth.

Economic goals can be woven into ecosystem restoration. In 1978, when Congress expanded Redwood National Park in California, most of the new parkland had been heavily logged. Between 1979 and 1987 the Park Service spent over $30 million on payroll, leases, and contracts to rehabilitate these damaged lands.[22] On a regional scale, if the Forest Service could secure funding for erosion control, retiring roads, and prescribed burning to reduce fuel loads and mimic natural fire patterns, an entirely new restoration economy of hundreds of millions of dollars would be created.

The scientific tasks called for by the biodiversity crisis may in time bring us to large-scale ecological restoration, returning wolves and grizzlies to California and Colorado and linking greater ecosystems across the country. If so, we will have become a different people. By 2090, humans will have burned almost all of the planet's currently known reserves of fossil fuels.[23] Douglas-firs 100 years old in the 1990s will have become part of an established pattern of ancient forest. With the wounds caused by resourcism healing, ecosystem management will have evolved into something new. This vision of a sustainable future inspires hope. It renews commitment to what most certainly will be decades of difficult work. In time, we may finally begin to experience what it means to be a part of, rather than the ruler of, nature. We, too, could become "native."

Conservation biology–based management will not by itself carry us to this far shore. This is the second point upon which a sustainable future depends. Robert Keiter is not the first to observe that "public lands policy simply cannot be defined solely in scientific terms. [People are] now a dominant force in nature, and human interests—economic, social, and political—must be accommodated."[24] This brand of pragmatism separates human interests from biodiversity concerns. But healthy economies rely on productive ecosystems. People are not disconnected from but rather related to the natural world, strained though the bond may be. Much of the strain has to do with human overpopulation, a problem that goes well beyond the scope of conservation biology. Keiter, however, is correct on one point. If ecosystem management is ever to provide a biodiversity protection network and a restorative future, the federal agencies, Congress, scientists, and citizens will be called upon to work together as they have never done before.

Land Managers in Support of Biodiversity

Just as a landscape evolves, so does land management. Both are complex composites of many forces. Unlike ecological change, however, management reflects human choices. The agencies know that they must respond to the unrelenting force of new scientific information. Conservation biology *will* be brought into management. At this early stage the agencies believe that their job is to "accommodate new forces while remaining as loyal as possible to the old."[25] Holding fast to the status quo, land managers choose those aspects of ecosystem management that will not rock the boat too violently, which results in a complex "stratigraphy of management," a bit of viable-population theory here and a new wildlife corridor there.

The question is how can ecosystem management, evolving within a social context of bureaucratic competition and elite, technocratic decision making, become integrated into the hearts, minds, and hands of land-management professionals? How can the turf-protecting tendencies of the Forest and Park Services be used in the service of biodiversity?

There are many ways to encourage enlightened federal land management. A first step would be to encourage land managers to understand that the biodiversity crisis is the result of social, economic, and political forces that go well beyond biology. Many managers (and biologists), according to policy analyst Tim Clark, retain the "unrealistic view that [ecosystem management is a] strictly technical biological task."[26] Yet there is much more to management than wildlife biology and ecology. To help land managers to better understand the range of forces that influence professional practice, Clark and his coworkers are developing a more broad-based framework that acknowledges the role of power relationships between individuals and agencies, bureaucratic behavior, political influences, and environmental values, as well as scientific study per se.[27]

With the competitive relationships of the past no longer serviceable in today's greenhouse world, this wholistic view is arriving none too soon. A case might once have been made for interagency rivalries producing political dividends, but that was before biodiversity, endangered species, ozone depletion, and atmospheric carbon became household words. Homo sapiens has appropriated from 20 to 40 percent of Earth's terrestrial primary production.[28] Our current environmental values, whether they apply to multiple-use or amenities, are untenable. The politics of timber versus preservation, for example, has been rendered superfluous by the current rate and scale of environmental deterioration. Organizational "party lines" hinder cooperation when problems expand across political boundaries and include numerous players. Competition only creates ineffective managers in the midst of the biodiversity crisis.

Managers are also ineffective because of the values reinforced by long-standing competitive relationships. Multiple-use is based on economics, not ecology: Its ideology is preecological and therefore inappropriate to today's land-management problems. What goals should a manager select? Conflicting legislative mandates range from bans on motorized equipment to funding for clearcut logging. Forest and Park Service staff often complain that there is no social consensus on protecting biodiversity. But conservation biology is now making clear that biodiversity is not just another multiple-use goal—it is a condition fundamental to viable ecosystems. Healthy parks are dependent on healthy forests and vice versa.

In the North Cascades, as elsewhere, biodiversity protection and ecosystem-level considerations are new to most Park and Forest Service managers. Education can provide a relatively nonthreatening forum for initiating agency cooperation. To begin, the agencies might engage a nearby university to host a conference or workshop on ecosystem management. The 1987 Transboundary Conference was a step in this direction. Initial meetings, however, need to become ongoing forums. Leadership does not have to originate with the unbending D.C. hierarchy. It could be ignited by a university faculty member, a gadfly district ranger, a sympathetic superintendent, or a regional forester. But the spark must spring from somewhere. Several biologists have pointed out the critical role of conservation biologists in land management. There are as yet few such scientists employed by the agencies.[29] This situation is beginning to change. The Forest Service, Bureau of Land Management, Fish and Wildlife Service, Environmental Protection Agency, and Park Service (belatedly) are presenting a short course on biodiversity for agency decision-makers. The Greater Ecosystem Alliance is sponsoring an ongoing public seminar series on ecosystem issues including wildlife, hydrology and cumulative effects, alternative silviculture, and more. Federal land managers and citizen activists are attending. For the long term, as funding becomes available, the Park and Forest Services (as well as other agencies) need to hire staff conservation biologists. In the meantime managers must become students of ecosystems if they are to become better stewards of the land.

While managers learn about conservation biology, they can also begin to cooperate on constructing regional scientific data bases. Mutual research, like education, is relatively nonthreatening. Gap analysis provides an excellent framework for agencies as they coordinate their efforts. The National Science Foundation-funded long-term ecological research program, which explores ecosystem trends over many decades, could serve as a model for mutual data collection.[30] The Forest Service's research natural area system, originally set up to protect representative examples of each timber type nationwide, should be redesigned and greatly enlarged to capture all types of national forest ecosystems. New areas, straddling park-forest borders, might be added to create an interagency research natural area network.

Groups such as the IGBC (Interagency Grizzly Bear Committee) can be useful in coordinating research, maintaining data, and determining research goals, but only if they are less politicized and more task-oriented will they prove effective in dealing with biodiversity problems. Any such committee should be composed of both independent and federal biologists, not just representatives of each interested agency. Ecologist Elliott Norse suggests that the agencies use a scientific oversight committee to provide substantive comment on forest and park plans. As it stands, input from the scientific community is sought only when there is great debate over a particular issue, and then it is not often followed. Because of the depth of the biodiversity crisis, agencies would do well to institutionalize nonpartisan scientific input at all stages of the planning process. A national coordinating body such as that proposed in the draft national-biological-diversity bill might serve as another model of coordination. Whatever advisory forums evolve, all will need to address regional landscape issues and perform adequate monitoring.

The Forest and Park Services should not limit themselves to in-house research and education. If Hal Salwasser is correct when he writes that "most societies wish to conserve species diversity," then citizens need information with which to set adequate environmental goals. In a democracy, as political scientist Bruce Jennings makes clear, if managers "are to cope, then they must be supported by an informed public."[31] Even as agencies work together to understand biodiversity, they must reach out through public education. In the Greater North Cascades, the Forest and National Park Services have taken the first steps along this path. Both agencies are funding, in part, an outreach coordinator who travels to local public schools to give presentations on the value of wildland ecosystems. In 1989 the National Park Service proclaimed biodiversity to be its nationwide interpretive theme for the future. While these developments are encouraging, outreach activities need to be expanded greatly. Agencies should give greater emphasis to the protection of biodiversity and the development of ecosystem management. Wilderness designation and commodity use should both be guided by the needs of long-term ecosystem protection. The ongoing campaign by agency biologists to replace Smokey Bear with the ecological story of fire could serve as a guide. But there is more to biodiversity education than fire ecology and balancing commodity uses. A Park Service directive describes the agencies' interpretive campaign as stemming from "a love of wild things and wild places and a continuing fascination with the mystery of how it all works."[32] This is what the revolution in land management is all about.

Research and education projects could serve to overcome the inertia with respect to ecosystem management that plagues the Forest and Park Services. These initial steps are less "risky," given the history of interagency conflict and the current political atmosphere. But biodiversity depends on more than research and education; ultimately its fate will be determined in the politically delicate areas of public participation, planning, and leadership.

The Forest Service has been criticized widely for its handling of the public participation mandates of the NEPA and the NFMA.[33] Such agency efforts have been thwarted by the land managers' view of themselves as a professional elite who have little need of public opinion. Agencies have handled public comment as a necessary evil to fulfill the requirements of law. Theirs has been a top-down, bureaucratic brand of planning labeled "rational-comprehensive," or "synoptic," by students of public policy-making. Synoptic planning, familiar to anyone who has read a forest or park plan, is "a systematic process in which objectives are formulated, alternatives are identified, choice criteria are applied to the alternatives, and a recommended course of action is identified."[34] Responding only to information already collected (and filtered), to alternatives already "preferred," agencies silence the voices of citizens. They attempt to hide explicit ethical issues behind the smokescreen of a values-neutral playing field, pretending that all views will be considered equally. Throw in planning issues rife with ethical conflict—"my job against some damn bird"—and there is no way to avoid adversity. Numerous studies have shown synoptic planning to be ineffective.[35] The fact that both the Forest and Park Services still cling to such technocratic methods is a direct measure of their desire to maintain decision-making control.

However one views this debate, I believe interagency cooperation in heeding public opinion would brighten the prospects for biodiversity. First, I would recommend a local-regional watershed approach rather than the traditional national forest or park focus. A pilot project in the Greater North Cascades might focus on the Cascade River or, on a broader scale, the Skagit River watershed. Land ownership in these areas is mixed (federal, state, private), but the checkerboard pattern that plagues interagency cooperation in many places in the West is not pronounced. Public meetings over a period of several months or longer would air biodiversity concerns and establish a common information base. Once trust and information were shared, specific planning could begin.

Agency attitudes toward citizen participation may be changing. The 1980s saw genuine participatory planning in the 3-million-acre Bob Marshall ecosystem in Montana. Known affectionately as the Bob, this may be the most famous wilderness in the country. By the late 1970s, thousands of hikers, hunters, and other visitors were eroding trails, cutting down trees for semipermanent camps, and doing general ecological damage. Early attempts to solve the problem found hunters and horsemen lined up against backpackers and botanists. All the groups were arrayed against the Forest Service. There was little prospect of resolution. Today the Bob is convalescing. It took a visionary manager, two separate planning models, and a time-consuming and fractious process, but consensus was reached and a management plan has been in operation since 1986.[36]

Gerry Stokes was the Forest Service manager who shepherded this process to completion. Stokes based his work on the Forest Service's excellent wilderness management model, the limits of acceptable change (LAC). LAC outlines a simple method that allows planners to set quantitative limits on wilderness use based on what levels of ecological change are acceptable. Change is measured against natural patterns in the landscape. The model can be just as ineffective as any other top-down planning method if citizens are neglected. Stokes's genius was to conceive of LAC as "a three-legged stool with a managerial leg, a research leg, and a citizens' leg." To build the citizens' leg of LAC, Stokes employed a decentralized, dialogue-based, transactive planning process.[37] He set up a task force of managers, planners, researchers, interest groups, and citizens from around the Bob to engage in small, ongoing group meetings. Issues were outlined and argued, research data was presented, assumptions were challenged. The goal was to "narrow the knowledge gap" between different parties and thereby set limits on use in the Bob that everyone could live with.

The Montana task force took three and a half long years to reach agreement. But with the new plan in effect, ecological conditions in the Bob have improved and members of the task force continue to stay involved, fine-tuning the original guidelines. Recently, in fact, the group called for "tailoring the management institutions to the land" by restructuring Forest Service administration to fit the ecological boundaries of the Bob.[38] With modification to reflect different management problems (biodiversity, new forestry) and regional ecosystem constraints, there is no reason why LAC-transactive planning should be limited to one wilderness area in the northern Rockies.

New decentralized planning methods could help to break down the walls that insulate managers from citizens and citizens from each other. Bruce Jennings points toward a broader, cultural view of public participation that complements what science is saying about land management: "If participation were seen in a broader, more positive way, as a process which has both instrumental and intrinsic value for those taking part and for society as a whole, then there is no reason, in principle, why the exercise of participatory (or representative) democratic governance need undermine the legitimate and valuable role played by experts or professionals in the policy making process. Democracy and elitism may be antithetical, but democracy and expertise are not."[39]

The Park and Forest Services have not yet embraced this view. But the Forest Service, under intense pressure due to the northern spotted owl, the threat to ancient forests, and nationwide overcutting, is attempting to respond. The agency's New Perspectives program has broad goals that appear similar to what I have been describing here as ecosystem management. New Perspectives incorporates ecosystem-scale planning, new forestry research and practice, conservation biology education and training for managers, and public education. According to the glossy brochure that describes the first pilot application of the program, the Shasta Costa Integrative Resource Project in the Siskiyou National Forest in southern Oregon, "New Perspectives invites people to become full partners in National Forest management. . . . It means building an alliance of people who can sit at the same table, agree to disagree, and not leave until they find solutions everyone can live with."[40]

Hal Salwasser is the national director of New Perspectives. (Recently, Salwasser left the Forest Service for a university teaching position.) He claims that the Forest Service is responding to changes in public attitudes about everything from producing board feet to protecting biodiversity.[41] Conservation biology is pushing the agency hard.

Unfortunately, as good as New Perspectives looks in a fancy full-color brochure, it may not be so revolutionary on the ground. It could be just another quick political fix. The Shasta Costa brochure and its accompanying draft EIS appear to be more style than substance. The publication begins with a very selective reading of history, claiming that the Forest Service has always based its management on the "bedrock" of Aldo Leopold's land ethic.[42] It goes on to present new forestry as if it were a tried and true method of doing business, not the long-term experiment that it really is. New Perspectives is characterized as using the "entire landscape to blend protection with production." But while there is much talk of new forestry techniques à la Franklin, there is no mention of the role of wilderness protection, the need to reduce the allowable cut, the exhausted condition of many front-country timber stands, or any of the other limiting factors that must be addressed *before* the program is put into practice. The idea of constraints on human use is absent from New Perspectives literature.

The Shasta Costa draft EIS, the first legal planning document to come out of New Perspectives, shows that the Forest Service still has room to improve. The agency proposes to "maintain biological diversity" in the mostly roadless Shasta Costa watershed by cutting 11 million board feet of timber over three years.[43] Using new forestry methods, there would be a 31 percent reduction of old-growth interior habitat by 1998. Since the Shasta Costa drainage lies between two of the few remaining wilderness areas in the region, this should be the last place to experiment with new forestry. Aside from the question of whether biodiversity might be better maintained by simply leaving the area alone, the Forest Service has also left the door open to more logging in Shasta Costa. Appendix F of the draft plan makes clear the agency's intent to get the cut out of Shasta Costa three years *after* the New Perspectives experiment is over so as to reach the Siskiyou National Forest's allowable cut level.

Critics of New Perspectives, like Jeff DeBonis, maintain that the Forest Service should be "asking whether or not we should be cutting, roading, mining, or grazing *at all* on public trust lands."[44] The managers responsible for the Shasta Costa plan believe that "the National Forests cannot, in practicality, be reserved from all timber harvest"; they see New Perspectives as "a method of compromise between protection and production."[45] The timber industry, on the other hand, wants New Perspectives to unlock the door to all the wilderness areas and national parks that have been set aside since 1970. Industry argues that if new forestry would better "preserve the ecological values" of reserves, then protected areas should be opened up to development.[46]

These are the poles between which hang the thread of life for native species and ecosystems in North America. New Perspectives will not contribute much to ecosystem management if the mistakes of the past are not acknowledged or the Forest Service continues its old balancing act of attempting to appease all (human) interest groups. There is little margin left if everyone is satisfied. A truly new perspective will not emerge until managers and citizens recognize the nexus of planning as human use nested in living ecosystems, not as park and forest islands in a sea of anthropogenic change. This vision would better serve *all* inhabitants of ecosystems—is in fact essential if we are to avoid the democratic trap of giving equal weight to the votes of those who, for short-term self-interest, would destroy biodiversity. The balance between use and protection, New Perspectives and native diversity, private and public interests, and democratic and elite decision making will not be achieved until we somehow learn to "give nature a vote" in our deliberations. And it is still politically untenable to broach the issue of incorporating nonhuman beings into ecological decision making. Watershed-based, decentralized ecosystem management for native diversity is not about how to sustain resourcism in its old or new (perspectives) forms; it is about providing a different kind of counsel that, over time, with plenty of hard work and lots of luck, may foster a sense of the basic limitations of being human in a finite world.

Passing an Endangered Ecosystems Act

"We can go on talking about environmental integrity until hell freezes over, but if Jim McClure [a past Republican senator from Idaho] wants us to produce 11 billion board feet, then, by God, we are going to produce 11 billion board feet," once said Steve Mealey, an assistant director of the Forest Service.[47]

How much logging is too much?

During the 1986 session of the Wyoming legislature, a bill was introduced that sought to disavow the very concept of a Greater Yellowstone ecosystem.[48]

Can regional ecosystems be voted out of existence?

Even as the Fish and Wildlife Service weighed the merits of the petition to list Columbia River salmonids under the Endangered Species Act, Senator Mark Hatfield (R-Oregon) was negotiating a "predecisional recovery plan" with a select group of business people. Few scientists were privy to the meetings, even though the Endangered Species Act calls for biology, not economics, to form the basis for recovery planning.

How many loopholes can Congress find in its handiwork?

Ecosystem management cannot be realized through conservation biology, the New Perspectives program, and improved interagency relationships alone. In theory, land management reform can thrive on process ecosystem management. But not without leadership. Like science, leadership is a matter of scale. Any district ranger or park superintendent can sponsor a day-long staff workshop on landscape ecology. An administrative agreement covering grizzly bear research in the Greater North Cascades would take regional-level cooperation between the Park and Forest Services. To pass a prescriptive biodiversity bill would require a concerted national campaign and congressional leadership. All these activities are based on a commitment to ecosystems and a willingness to take risks. There are limits, however, to administrative reform, and the buck stops on the marble steps of Capitol Hill.

Congress is not quite ready to give genes, grizzlies, and greater ecosystems full support. When it comes to biodiversity, legislators dance a modified version of the tango, two paces forward and several steps back. Lawmakers are moving as fast as they can to stay in the same place. In 1990 and 1991, Congress defeated efforts to gut the Endangered Species Act. Yet Jim Jontz's ancient-forest protection act never received a full committee hearing. The Senate rejected Mark Hatfield's timber industry bailout bill, with its high logging levels, only to sign a 1991 Forest Service appropriations bill that directed the agency to cut down forests in the Pacific Northwest at least 60 percent faster than they can grow back.[49] Those national forests that met these top-down timber targets received a 5 percent budget boost for wildlife, watershed, and recreation projects. Ecosystems are protected with the money generated by their destruction.

Congress, therefore, must be willing to break ranks with business interests and afford endangered ecosystems top priority. There are numerous models available for lawmakers to follow. The most obvious example is the Endangered Species Act. There is nothing in the act about "protection being economically feasible. . . . [In fact, economics] . . . appears counter to the list [of values to be protected]."[50] Yet there is a great difference between working with a radical law on the books since 1973 and passing new protective legislation today. The difference between the 1970s and the 1990s is not in the political climate. Politics is still about power and compromise. The crucial distinction lies in how much more aware we are of the fundamental importance of functional ecosystems and of the mounting local, regional, and global cost of their destruction. In passing the Endangered Species Act Congress did not say, Save only those species that are economically valuable. It said, Protect all species regardless of the cost. Today, mostly because of ineffective implementation, endangered species' problems have become endangered ecosystems' dilemmas. Congress will soon have to respond in kind. Lawmakers must consider the kind of legislation recently proposed in the preamble of a model bill drafted by grassroots activists:

Section 2 ENDANGERED NATIVE ECOSYSTEMS

(a) The Congress finds and declares that—

(1) the survival of myriad species in the wild depends on a well-functioning native ecosystem whose members live together in complex and interlocking ways, such that the fate of each function and each species is closely tied to the fate of others;

(2) the only feasible method to maintain function and diversity over the long term and thereby avert increasingly common and hopeless extinction crises for the many species whose life re-

> quirements have not yet been fully determined, is the recognition and protection of entire optimum dynamic areas for each native ecosystem;
> (3) various native ecosystems in the United States have been rendered extinct as a consequence of economic growth and development untempered by adequate concern and conservation;
> (4) other native ecosystems have been so depleted in area that they are in danger of, or threatened with, extinction.[51]

This sort of bill could mean the difference between life and death for thousands of North American plants and animals over the coming decades. For humans, it might also determine whether they have healthy air to breathe, clean water to drink, and wildlands to experience. Scientists and environmentalists, radicals and reformers alike, are calling for an endangered ecosystem act.[52] The federal Office of Technological Assessment maintains that protecting ecosystems is the only way to ensure that species will continue to evolve.

An endangered ecosystems act might mandate gap-analysis research to identify ecosystem types at risk. A biodiversity-protection network could be built into the new legislation. Another approach might be to identify and protect endangered ecosystem fragments while bringing representative examples of all U.S. ecosystem types into a reserve network. Activists in Montana and Idaho have drawn up a Northern Rockies Ecosystem Protection Act that identifies five greater ecosystems across 35 million acres and the biological corridors that, with restoration, could exist between them.[53] And there also exists a draft bill that focuses on protecting ancient forests nationwide, not just those in the Pacific Northwest.

Whatever the legislative approach, there will need to be restrictions on human activities that threaten ecosystem structures and functions. The movement of species in response to the projected effects of global climate change should be accounted for, with biological corridors a major part of any law. Long-term restoration of damaged lands could be facilitated by the recreation of a civilian conservation corps employed to plant trees, close roads, and bring back anadromous fish populations. Whatever its exact provisions, the new law cannot be just a national parks act updated to fit the twenty-first century. The whole notion of multiple-use lands and reserves is anachronistic at best when animals roam across large landscapes, air pollution travels hundreds of miles, and so much land needs restoration.

But if the state of draft ancient forest legislation is any indication, Congress will find it an onerous task to pass an endangered ecosystem act. A law focusing on large national parks might be less risky politically and a pump-priming step in the right direction.

A national parks biodiversity-study act would mandate a two-year examination of all parks over 50,000 acres. Congress could enjoin the Society for Conservation Biology or the Ecological Society of America, or convene a special working committee of the National Academy of Sciences, to complete the study. The research would determine the ability of each park to protect and maintain functional internal ecosystems as well as examine its role in the larger regional landscape; assess how each park might contribute ecosystem types and biodiversity hotspots to a biodiversity-protection network; evaluate the impacts on parkland biodiversity from development on surrounding federal, state, and private lands; and predict the population-viability status over the next fifty years of all species native to the parks. The public report would be an ecological snapshot of current conditions in the most treasured of the nation's natural areas. It would likely startle, then anger, millions of people and catalyze enough pressure to get Congress moving on an endangered ecosystem act.

Congress is unlikely to act for ecosystems without first feeling political pressure, and the loss of biodiversity, like global warming, may not generate that pressure quickly enough to bring about timely action. No member of Congress will do anything without information. Providing the details is the job of citizens, scientists, managers, and all those who would speak for nature. Outlining a bill grounded in conservation biology may seem utopian, but it would help to bring the message to Capitol Hill. That, however, is only the first step. The second step, as Franklin D. Roosevelt once reminded a reform delegation, is more tricky: "Okay, you've convinced me. Now go out and bring pressure on me."

Protecting Biodiversity on State and Private Lands

Whatever forms of political pressure activists adopt, they must not focus exclusively on the public lands.[54] Difficult as initiating ecosystem management on federal lands promises to be, the public domain, by itself, is simply not large enough to protect biodiversity. This is the political puzzle at the core of the crisis. The plight of the Florida panther proves the point. In south Florida, 45 percent of the panther's range lies on private land.[55] If the current rapid pace of habitat destruction continues and the cat is eventually limited to state and federal lands, its population, according to estimates, will range from nine to twenty-two animals. Even in the relatively pristine 8-million-acre Greater North Cascades ecosystem, the grizzly bear and the gray wolf need more living room. Ecosystem management restricted to government lands is a prescription for extinction. Impossible as it may be to imagine, management should serve private- as well as public-land panthers and state- as well as federal-land grizzlies and wolves. The question is how?

At the national level, the passage of an endangered ecosystem act covering all jurisdictions would show Congress's intent to lead on this issue. The NEPA could be amended to include impacts on biodiversity in all EISs. (If passed, the draft national biological-diversity bill would do just that.) A crash funding program for gap analysis, interagency research, and National Science Foundation conservation-biology grants would go a long way toward saving what Aldo Leopold called "all the cogs and wheels" of native diversity.

Across regional ecosystems containing federal, state, and private lands, management will depend on shared organizational structures, scientific definitions and data, information, and knowledge of the cumulative effects of individual development proposals. No longer can the right hand of the state move without the left hand of the Park and Forest Services.

Federal agencies like the Forest Service might share biodiversity information more readily if their organizational structures shifted focus from resources (timber, wildlife, recreation) to ecosystems (ancient-forest managers, riparian rangers, ponderosa-pine planners). The entire state and federal land-management bureaucracy would derive long-term benefits from integration: state departments of parks, natural resources, and wildlife merged into national parks and forests according to landscape-level ecosystem boundaries. (The current proposal from the LAC task force in the Bob Marshall ecosystem is a bold step in this direction.)

Managers' inability to cope with cumulative effects is hastening the trend toward an ecosystemic blending of federal, state, and private bailiwicks. Both the NEPA and the NFMA, as well as most state environmental-policy acts, require reviewers to evaluate a specific road construction proposal, housing development, or timber sale in light of "the number of similar projects going on in the region, the likelihood of repeated projects of the same type, and . . . proposed actions in the vicinity [that] might compound the projects' effects on the quality of the environment."[56] Within the current balkanized framework, however, cumulative effects constitute the black hole of environmental-impact analysis. Managers have little incentive to look beyond their own jurisdictions.

The legal wording of state cumulative-effects provisions makes the NFMA seem like a model of legislative clarity. Under Washington State's Environmental Quality Act, "common sense is the best guide" to evaluating cumulative effects and planners are directed to "make reasonable reference to past projects and future expectations."[57] In Wisconsin and California, state laws prescribe no specific methods for completing any analysis.[58] On state and private lands in California, environmental reviews are the responsibility of the developer rather than the government, a sort of fox-guarding-the-henhouse arrangement.

When ownership patterns become mixed, as they are throughout most of the United States, the cracks in determining cumulative effects widen to chasms. I have read hundreds of timber-sale impact assessments from the northern half of the Mount Baker–Snoqualmie National Forest. Each one reaches the same conclusion: "There will be no significant impact on the grizzly bear due to this project." But neither the Forest Service nor any other review agency tracks the amount of loss in a given type of ecosystem over time. At some point, when enough sales are allowed in grizzly bear habitat across the Greater North Cascades, the bear suffers. Or the northern spotted owl, or the pine marten, or the Vaux's swift, the cougar, the northern flying squirrel, the dusky shrew, the tailed frog . . .

Adequate cumulative-effects analysis depends on two things: better knowledge of species and ecosystems, and more sophisticated planning that marries different agency processes. Gap analysis and interagency inventories of biodiversity will respond to the first problem. To begin to solve the second issue, federal and state laws will need to be amended to define more specifically the procedures of cumulative-effects analysis; to set up a federal-state-private information clearinghouse and habitat-loss tracking system based on regional ecosystem types; and to create a tax disincentive and/or land purchasing fund for private-owner compensation.

The first hesitant steps in breaking down state-level administrative barriers to ecosystem management are already being taken. A joint University of California–state government report, *In Our Own Hands: A Strategy for Conserving Biological Diversity in California,* was published in 1990.[59] It recommended a coordinated approach to biodiversity protection, including passage of a California habitat-protection act, establishment of a biodiversity research institute, closing loopholes in current environmental laws, and implementing a full environmental-education public school program.

In 1991, the California State Assembly Office of Research published *California 2000: Biological Ghettos,* which offered a sobering preview of environmental quality in the state over the next thirty years.[60] With ongoing political conflicts over logging restrictions, urban sprawl, water availability, and endangered species, the state decided to act. On September 19, 1991, California became the first state in the country to adopt a comprehensive strategy to protect regional biological diversity. The leaders of ten state and federal land-management agencies signed a Memorandum of Understanding to "work cooperatively in conserving biodiversity on a regional basis across administrative boundaries."[61] Parties to the agreement include the secretary of the California Resources Agency; the state departments of fish and game, forestry, and parks; the U.S. Forest Service; the National Park Service; the Bureau of Land Management; and the University of California, among others.

The structure of the agreement is straightforward. It creates an executive council on biological diversity composed of the heads of the signatory agencies. The council will set policy and design statewide strategy across all ownerships. Various special interest groups will participate in the council's work but will have no vote. At the regional level around the state, bioregional councils will convene to implement the policies of the executive council. And, within regions, watershed and landscape associations will serve a similar function.

This policy development is a step in the right direction. But there are several problems with the biodiversity strategy. It was adopted for economic as well as ecological reasons, and it is unclear from the mission statement which of these is the state's primary concern. As is the case with most interagency agreements, the executive council has no legal authority to enforce its directives, nor is funding provided for its activities. The council provides no voting membership to any group outside of government. Policy research elsewhere has shown time and again that this is not a prescription for changing the status quo.[62] Some have even suggested that agency bureaucrats have merely appropriated the language of reform (biodiversity, bioregions, watersheds, ecosystem management) and will continue to favor development over ecosystem protection.

Washington state, embroiled as it is in the ancient-forest controversy, is ahead of most states in seeking ways to blend private, state, and federal land management. Brian Boyle, past state commissioner of public lands, succeeded in bringing state and private interests groups together by engineering several consensus-based task forces. In 1986, a timber-fish-wildlife agreement brought some minor reforms to a very weak Washington Forest Practices Act. Two years later, faced with the impending loss of all old-growth on state forestland, Boyle convened the Committee on Old Growth Alternatives. This group included representatives of environmentalists, the timber industry, Indian tribes, and legislators, and various education, policy, legal, and scientific experts. In 1989 the committee opted for a large-scale experiment instead of ecosystem management with a plan for the 260,000-acre

Hoh/Clearwater Block on the Olympic Peninsula, the largest remaining state old-growth forest. Two-thirds of the old-growth would be logged, the entire area would be put under new forestry management, and economic development and retraining monies would be made available to local communities.[63]

Beginning in 1989 a third task force, the Sustainable Forestry Roundtable, wrestled with the rate of logging on private timberland and a new round of proposed reforms in the Forest Practices Act. In 1990 tentative agreement was reached about establishing greater erosion control and an early-warning rule under which any landowner planning to cut more than 4 percent of a watershed in a year has to submit to a special state review.

Although overcutting was the issue that brought the roundtable together, no direct limits on logging could be agreed upon. Considered instead was a "10 percent for wildlife" initiative. This new rule would have required the State Department of Wildlife to select 1 million acres, a tenth of Washington's state and private timber base, to be protected as old-growth habitat.[64] Since most state forestland is now second growth, the rule would have been a revolutionary prescription for growing new ancient forests.

The members of the Sustainable Forestry Roundtable worked together under pressure, used mediation techniques, and agreed to disagree. But to little avail. They could not reach consensus and a legislative attempt to revive the proposal failed. There were good reasons for this: The Roundtable never prescribed sustainable forestry. Cutting rates under the agreement would have been roughly twice the level of sustained yield. The 10 percent for wildlife provision was based on expediency, not ecology—most of the acres would have been selected from riparian ecosystems that, because of water-quality concerns, already enjoyed a high level of protection. The initiative was simply the best deal environmentalists could cut with the timber industry. It certainly was not the "major conceptual shift from monoculture to landscape ecology" that Department of Wildlife director Curt Smitch trumpeted to the media. Conservation biology is based on scientific principles extended across entire landscapes. It is not limited to riparian corridors or some percentage of the commercial timber base.

As the irresistible force of sustainable ecosystem management prods the immovable mountain of entrenched political and economic interests, the thorniest issues surrounding landscape-level management are left untouched. The balance between national and local control and the belief in private ownership go down to the bedrock of American culture. These two issues will loom larger in the biodiversity crisis as time goes by and the noose of extinction and habitat fragmentation tightens around native diversity. I cannot resolve them here, but I can make several observations.

Dave Foreman, cofounder of Earth First! and former D.C. lobbyist for the Wilderness Society, reminds us that there would be precious little legal wilderness if it were up to rural residents, county commissioners, and local and state governments.[65] The national interest has created the current reserve system, and the public will play a major role in making it stronger for the immediate future. On the other hand, there can be no lasting ecosystem management, no land-management revolution, unless we step back from the sort of centralized, bureaucratic decision making that emanates from inside the Washington, D.C., beltway. North Cascade National Park would not exist but for congressional action, yet the park's general-management plan, written by agency planners in Denver, looks more to the past than to the future. For now, the grizzly bear needs both the federal government and grassroots pressure groups. Later, its future may depend more on local and regional decision-makers. Choices about how humans live with ecosystems can only be, as Wendell Berry points out, specific to local places, conditions, and practices.[66] What works for people and the bear in the Greater North Cascades will be different from, even as it should be coordinated with, what works in Greater Yellowstone. The paradox of national versus local control will evolve as conservation biology and ecosystem management reframe our experiences with these concepts.

The concept of private property will also change as the biodiversity crisis deepens. Practical strategies to bring private lands immediately into ecosystem management include (in order of likelihood of short-term success) zoning law changes, outright purchase and/or condemnation, leasing agreements, tax incentives, conservation easements, land trusts, stewardship programs, and education. A 1986 study of private landowners whose property harbored important habitat for wildlife found that if they were simply made aware of the biological value of their land, this alone would induce 20 percent of them to leave the land undeveloped.[67] A more recent study has revealed that landowners support a mix of education, tax breaks, and stewardship programs that recognize their work on behalf of wildlife over other management methods.[68] These findings are heartening. They indicate that the grip of private property and its attendant profits may not be as strong as it appears. Many private owners may be ignorant about biodiversity now, but with education and encouragement they could soon be inclined toward its protection.

Political movements like Marxism and socialism once challenged our collective image of nature as private property and commodity. Ecosystem management is based in biology, not economics. This is a critical distinction. A vote against ecosystems is not a victory for one party over another; it represents another increment in the loss of life support for both sides. Supporters of biodiversity must not forget that Washington, D.C., North Cascades National Park, and Okanogan National Forest are artificial boundaries, nonexistent for pine martens and red cedar trees. Few species will survive on any kind of land without protection until biology pushes our values away from prohibitions and toward the affirmation of all life. Because Congress is ill-prepared to offer this brand of leadership, it will have to spring from somewhere else.

Scientists and Advocacy

It is raining and I am deep in Douglas-fir ancient forest up the Baker River, looking for a suitable classroom. The rain is gentle, almost mist. Swatches of blue sky tell me that the shower will soon pass, a brief moment of moisture spun out of a summer mountain afternoon. Long sun rays shaft through the multilayered canopy and I discover a giant down log perched above vine maple and ferns with a niche underneath just right for a group of thirteen students. I sit down in the deep moss and duff, lean against a red cedar, and consider how the place might illuminate my lesson plan.

Tomorrow is the final exam, after which we will hike out to the trailhead and drive back to Bellingham to end our three-week field exploration of ecosystem management in the Greater North Cascades. It has been a rich month of traversing mountains from east to west. I have several loose ends to tie up before the test. This place, out of sight of the trail, will help me with my work. For an in-class exercise, the students map out the distribution and age classes of the down logs in the immediate vicinity. All along we have been challenging our bias toward standing live trees by investigating the unseen work of the dead and down. We have dug soil pits and peeled off bark to search for invertebrates and fungi, and this afternoon we will map log locations, look for patterns, and estimate the amount of decaying wood nearby. For the second half of the class we will sit under the cedar and discuss the question What is natural? based on several articles in our course reader. Is this low-elevation valley forest natural, cut off as it is from other old-growth stands? Are our nylon tents and aluminum backpacks civilized artifacts or tokens of just another species of mammal? Is the foot trail back to camp as sustainable as the creosote-treated timber and metal-railed footbridge spanning the river?

We have found few answers to these questions. What we have discovered is the need for a genuine reframing of our image of and attitude toward wild nature. After miles of trails, hours of meetings, and endless discussions in class and around the campfire, I have come to the conclusion that teaching without direct engagement is the same as managing land from behind a desk or planning from an abstract model. The best way to protect biodiversity is to invite students and managers and citizens and lawmakers out of classrooms and offices and into ancient forests.

Who are the people most qualified to teach ecology? Biologists who study natural history and ecosystems and know them firsthand need to strengthen their commitment to sharing knowledge in forests and courtrooms as well as in classrooms. "It is only the scholar," Aldo Leopold wrote in 1949, "who understands why the raw wilderness gives definition and meaning to the human enterprise."[69] If Leopold's observation remains true today, then we are in deep trouble; faced with unprecedented deterioration of the planet's life-support systems, we should now know that wild ecosystems provide more than meaning.

Scientists must teach, speak out, and advocate for biodiversity. But there is a significant barrier between practicing science and practicing activism. Biologists have been reluctant to stand up for nature publicly because of the chance they might violate professional standards of objectivity and thereby lose esteem. They are not trained to communicate scientific concepts in lay language or to use the mass media effectively, and they often appear confused over the difference between interpreting facts and expressing values.[70] Scientists are shaped by their culture as well as by their profession to seek technical solutions to problems, regardless of their ethical content.

In university classrooms, field ecology and natural history are going the way of the passenger pigeon. Biologist Jack Carter and his coauthors, after surveying the state of the American curriculum, remarked, "If biologists are uninterested in acquainting themselves and their undergraduate students with the natural world that controls the destiny of life on Earth, the value of biology departments to education in the liberal arts stands in question."[71] Too many teachers, managers, and researchers are trapped by the Western positivist image of science as value-free; this, combined with the competitive system of professional rewards and incentives, has produced teachers who value research over education and researchers who care more about the credibility of their results than the protection of their study sites.

Biologists might realize that science, like everything else, is shot through with values. Sorting out the norms behind positions is the initial step of critical thinking. Researchers need to focus on what solutions are supported by the best available data. Their job is not merely to support what is "socially acceptable"; it is to help citizens to think more clearly about the social goals of environmental management.

Professional groups like the Society of Conservation Biology are in a perfect position to evaluate the scientific method behind Forest and Park Service management proposals. To improve scientific methods and enhance the credibility of the agencies, public planning documents should be critically reviewed. Every ecological society could stake out a piece of ground—there are more than enough plans to go around. The Northwest chapter of the American Fisheries Society has done admirable work in advocating for wild salmon; other regional chapters might become active as well. The prestigious Ecological Society of America could have working groups review the science found in management plans, not just when invited but as a matter of course. Scientists would then have an impact on the application of their research and agencies would spend less time in court.

Citizens and Congress would benefit from an ecological education. Biologists have a wealth of knowledge to share. One of the most important lessons for lay people has to do with what Paul Ehrlich calls the ability to see beyond short-term conditions.[72] One of my most effective teaching exercises has been mapping the down logs in the Baker River ancient forest. Students realize that the log they are sitting on not only lived upright for 800 years but will continue to make a vital contribution for at least 300 more years as it feeds a linear patch of salamanders, saplings, and duff. Such lessons are not available in second-growth stands or cities. To connect different scales of biodiversity, gene pools and landscapes, for example, helps "people make sense of the world around them. It asks them the most basic questions: What is this? Where am I? and then penetrates deeper into the questions that connect us with all beings: Who are you? Who am I? How do we fit together in the world?"[73]

Teaching biologists are needed at all levels of society, from the grade-school classroom to the halls of Congress. In Florida, Ecologists for Education in Local Natural History, an informal group of professional academics and educators, is bringing ecosystem awareness into elementary schools.[74] In California, a conservation-biology high-school curriculum is being tested in public classrooms.[75] And, for those biologists who wish to tackle Congress, there is an informal field guide to lobbying.[76]

There are students aplenty to teach. Many people in the urban United States are unaware that ecosystems even exist. Like FDR, my students may be convinced but they have not yet felt the pressure to change. The number of students taught or the force of facts alone will no more resolve the biodiversity crisis than viable-population research or an endangered ecosystems act. Ecosystem management works like a set of Chinese boxes: The box I label scientific advocacy and education fits into a larger box called public participation. In times of crisis the end result of ecological education must be a people aroused to action.

Biodiversity and Civic Responsibility

Who are the people?

The final plank of an interim ecosystem-management bridge can only be constructed by citizens willing to become intimately involved with their earth households.

The full range of environmental activism in the United States is practiced in the Greater North Cascades. There is the Sierra Club, whose 500,000 national members have a per capita average income of over $40,000 per year. The club focuses its considerable political power on Washington, D.C., where it has controlled the ancient-forest lobby. No old-growth protection bill could ever pass Congress without first being approved by the Sierra Club. The National Audubon Society holds the other key vote on ancient forests. Brock Evans, the Audubon lobbyist is completely committed to old-growth. Local grass-roots Audubon chapters are also very active in the Greater North Cascades. They were instrumental in organizing the Ancient Forest Alliance, a fragile coalition of activists pushing for protection. The alliance was augmented in 1991 by the Western Ancient Forests Campaign, which, through foundation support, allowed coalition members to open a Washington, D.C., office to lobby and coordinate the passage of a new law. Still, grass-roots groups do not have the political influence of the relatively conservative Sierra Club.

The Wilderness Society is the research arm of the environmental movement in the Pacific Northwest. In 1991, the society's Peter Morrison completed the most up-to-date scientific old-growth inventory of the Region 6 national forests.[77] The Scientific Panel on Old Growth went to Morrison for data. This research affects policy-making in D.C. only indirectly—the Wilderness Society does not have the lobbying clout of the Sierra Club or Audubon Society.

The Greater Ecosystem Alliance represents a new breed of grassroots activism that has risen in the North Cascades to fight for transboundary issues. The alliance is a classic, small-scale nonprofit organization with a full-time staff of three. Other citizens' groups, like FOCUS, (Forest Concerns of the Upper Skagit), have formed to generate logging policy reform at the watershed scale. A network of volunteer ad hoc groups, SUSTAIN, has organized to change statewide timber practices.

If the Wilderness Society represents the research arm of ancient-forest environmentalism, the Earth First! movement is the strong arm of the fight for biodiversity protection. Through guerilla theater at federal timber auctions—chaining themselves to bulldozers and skidders at logging shows and spiking planned timber sales in grizzly bear habitat—Earth First!ers seek cessation of business-as-usual. This group of radicals may be the most effective of all. Earth First! has led the way in shifting the U.S. environmental movement away from a national park/wilderness/aesthetic/recreation focus toward a biodiversity/ecosystems perspective.[78]

This is the landscape of activism in the Greater North Cascades. The timber industry is also organized for grassroots participation. Many Indian tribes also have an active voice in regional ecosystem issues. There are local interest groups ranging from hunters and hikers to off-road vehicle enthusiasts. Yet, even with all these avenues open to activists, most citizens in the Greater North Cascades, the Pacific Northwest, and the country as a whole do not participate directly in environmental politics.

There are numerous barriers to participation in public affairs in the United States.[79] Political philosopher Benjamin Barber believes that representative democracy, where some of the people, chosen by those who vote, govern all public affairs all of the time, needs to be revitalized by a strong democracy where all of the people govern themselves in at least some public matters at least some of the time.[80] Representation, coupled with government bureaucracy, has stripped citizens of civic responsibility. The average person has little access to policymakers, the boundaries between national and local interests are fuzzy, time constraints make it difficult to juggle an active public life and a private one, the participatory process seems inefficient, and complex issues put the spotlight on professional rather than lay opinion.

The northern spotted owl issue illustrates these points. As biologists built an overwhelming case in support of the owl in the mid 1980s, the Sierra Club, the Wilderness Society, and other mainstream environmental groups were hesitant to petition the Fish and Wildlife Service for a listing under the Endangered Species Act. The Sierra Club was convinced that to do so would invite an attack on the act by the Reagan administration and a prodevelopment Congress. Old-growth forests were not yet a national issue, and environmental professionals in D.C. were not ready to orchestrate a national campaign. All this was counter to what Northwest grassroots groups thought was in the best interests of the bird. Local Audubon chapters in Washington and Oregon and the fledgling Greater Ecosystem Alliance demanded immediate action, but they had little clout with lobbyists and no time or money to travel to Washington and apply pressure of their own. Not part of the D.C. power structure, their efforts would have been discounted anyway. The impasse was broken when Greenworld, an unknown group from Massachusetts, sent a listing petition to the Fish and Wildlife Service.

Environmental historian Donald Worster writes that "the promotion of democracy, defined as the dispersal of power into as many hands as possible, is a direct and necessary, though perhaps not sufficient, means to achieve ecological stability."[81] Many hands are already joining across the Greater North Cascades to promote such stability. Led by the example set by Washington Earth First!, local grassroots activists have learned the monkeywrenching technique of timber sales appeals. These slow the pace of cutting by bogging down the bureaucracy in time-consuming documentation. The trees breathe for another week, and a cougar slips through a corridor that otherwise might have disappeared. Appeals are the Forest Service's major complaint against environmentalists and the roots of an attempt by that agency, the timber industry, and some members of Congress to limit citizen use of administrative and legal channels.

In March 1992, the Forest Service announced that it would seek to eliminate citizen appeals of administrative decisions such as timber sales, oil and gas leases, and grazing permits. The agency claimed that the current process was creating "uncertainty as to the Forest Service's ability to deliver . . . goods and services, impeding economic growth and development."[82] This claim was contradicted, however, by a federal Office of Technology Assessment study which found that appeals had "not significantly decreased the volume of timber for sale," and that "appeals play a significant role in exposing inadequate environmental analyses."[83]

Grassroots activism has not been limited to appeals and the courts. Following the advice of a volunteer committee to increase public awareness of the state of the region's water supply, Whatcom County and the City of Bellingham, with funding from the Washington State Department of Ecology, began a program called "We All Live Downstream." In the Skagit River valley, FOCUS is working on sustainable forestry that combines economic, recreational, and wildlife concerns on Sauk Mountain, the sacred mountain of the middle Skagit basin. FOCUS is a member of the SUSTAIN coalition, which is drafting a state forestry initiative. Northwest Audubon chapters have pioneered the Adopt-a-Forest program under which volunteers map, research, and lobby for individual old-growth stands. This allows people to contribute to statewide ancient-forest mapping through direct encounter with down logs, snags, and centuries-old trees. Adopt-a-Wilderness and Adopt-a-Timber Sale programs are in place throughout the West.[84] Such programs, by encouraging a more informed citizenry, help to increase biodiversity protection.

Across the border in British Columbia, citizens have more to lose and fewer tools with which to slow down ecosystem destruction. Low-elevation forests are more fragmented there than in the United States, and the Canadian legal system provides little standing to those who would sue for biodiversity. But change is in the wind. In 1991, the University of Victoria hosted the Community Forestry Conference on Vancouver Island. Conference participants identified a framework for local control of forestry decisions and environmentally sensitive logging practices that "more clearly reflect long-term community values." That same year the Canadian-U.S. Ecoforestry Institute was founded to work on privately funded research in new forestry and sustainable woodlands management.

Beyond the Greater Northern Cascades, people are standing up and speaking out for their local ecosystems. The STP (Save the Planet) movement is organizing wherever people see the need to build networks across boundaries that have kept environmentalists, native peoples, inner-city social activists, and other interest groups apart.[85] In Illinois, a network of over 2,000 volunteers is spending weekends restoring two statewide endangered ecosystems, tallgrass prairie and oak savannah. Parents with schoolchildren are having to respond to questions like Where does our garbage go? and Why do we leave the lights on? As teachers are discovering, children are powerful agents of change.[86]

The scientific journal *Conservation Biology* recently reported on a participatory wildlife-management project in Zambia.[87] When villagers were hired as wildlife scouts and custodians, park policy decisions brought before village councils, and monies from tourism redirected to villages from the local national park, conflicts between park staff and local residents decreased, government costs went down, and better management resulted. I have visited one of the most inspiring grassroots management efforts, the Community Baboon Sanctuary, located in and around the village of Bermudian Landing in north central Belize. The sanctuary was organized to protect black howler monkeys and provide cash income to townspeople.[88] In 1986 Rob Horwich, an American primatologist, went directly to the village council with his idea, and the sanctuary was set up without any government or organizational support. To this day it is run by the villagers on a volunteer basis. The howlers are thriving. In 1991, 5,000 people visited the village to learn about one of the largest of the New World's endangered primates.

Volunteer grassroots activism has attained the status of policy at the Community Baboon Sanctuary. The LAC process in the Bob Marshall ecosystem proves that this can happen on U.S. public lands as well. But in the Greater North Cascades no such transformation has occurred. While the black howler monkey benefits from community efforts, the northern spotted owl is supported by some, vilified by others. Part of the problem has to do with scale. It is much easier to reach agreement over a handful of problems in a small Central American village or in a lightly populated area of Montana, where most commercial interests depend on wildlands protection, than it is in northern spotted owl territory. Even within the environmental community of the Northwest—conflicts with loggers aside—there is a diversity of views over how to protect ancient forests.

The recent history of U.S. environmentalism reveals several trends in the environmental movement: increasing memberships, more sophisticated research, the rise of professional bureaucracies, and a focus on lobbying in Washington, D.C. The largest groups, formed from local and regional chapters and clubs, have parlayed themselves into national power players. With growth and corporate-style leadership from business and government, they have shown a greater reluctance to challenge the political and economic status quo, and there now exists a "class difference" between national and local environmentalists.[89] After the Vision Yellowstone fiasco, Robert Barbee, superintendent of Yellowstone National Park, and others summarized their experience by noting that many conservation groups were "not there for the fight"; "to this day, we don't really know what the largest groups at their national offices thought of the *Vision*."[90] The top-down structure of large organizations does not serve owls and ecosystems any better than it does the Forest and Park Services.

Nor do hierarchical structures serve grassroots activists and the local people most affected by land-management policies. When citizen participation is politicized into influence groups, three classes emerge: those who lobby and are lobbied (Congress, the Sierra Club, the National Audubon Society, etc.), those who are listened to but discounted (grassroots environmentalists), and those who do not have a voice or vote at all (grizzly bears and Douglas-firs). Mainstream environmental groups do not often acknowledge the differences between and within the first two of these groups in values, knowledge of the issues, and political power and strategy.[91] But these differences are crucial to the success of a participatory movement for ecosystem protection for at least two reasons. First, ecosystem-management goals, as presented here, require such massive social transitions that new alliances must be forged to swell the ranks of those who want to be part of a sustainable future. Second, regardless of the leap of faith involved, there can be no hope of breaking down the class structure of present-day environmental politics unless people of many persuasions are allowed a voice in setting the agenda. "Involving local activists and naturalists in scoping and ground checking [ecosystem protection plans] is not the same," remarks grassroots organizer Richard Grossman, "as intentionally setting out to reach people in order to learn their views, to seek their participation in framing the problems and finding solutions, and to talk about the economic and political power which is destroying the forests. . . . It is about building trust with people who may now be hostile to 'environmentalists,' but who *also* believe that protection of [ecosystems] from corporate exploiters is essential. . . ."[92]

Working within the goals of ecosystem management for native diversity, people must be allowed to find ways to take care of the forest even as they make their living from it. There is certain risk here: How can people who don't understand ecosystems, who have participated at times in their destruction, seek to protect them? I believe this hazard is balanced by another political fact of life: Those who remain disenfranchised by strategies to sustain nature will not likely act in support of ecosystem management. Until the onset of the biodiversity crisis, our political system produced enough token environmental reform to satisfy most people. Compromise could almost always be reached. Now, from the northern spotted owl of the Greater North Cascades to the carbon-loaded atmosphere of Earth, the news is compelling: "Ecological systems have a dignity, not a price."[93] Ecosystem management offers two minimum criteria for ecological dignity: the continued evolution of species, and the functioning of ecosystems. This is nature's bottom line. Congress, land managers, and citizens have always been able to choose between mandatory environmental policies and those that reflect public preference. The biodiversity crisis has ushered in a world where the word *must* is becoming the rule of management. The rise of grassroots participation is the public's response to the substitution of *must* for *may*, and so far a thin majority of people seem to be siding with the grizzly. Dave Foreman describes the rise of citizen involvement as "the new conservation movement," where local and regional groups base their activism in conservation biology, an ethical obligation to ecosystems, and independence from national groups in Washington, D.C.[94]

It is still too early in the evolution from wise-use to ecosystems to gauge whether the social costs and inevitable dislocations will eventually overwhelm the very people who are now clamoring for grizzlies and greater ecosystems. Ecosystem management for native diversity, as I have outlined it here, is a complex set of forces: conservation biology, gap analysis, a biodiversity-protection network, new forestry, restoration of damaged lands, interagency cooperation, local and regional planning, an endangered ecosystems act, administrative partnerships, scientific advocacy and education, and grassroots activism. And these are only part of a more comprehensive force, culture at large. The connection among these parts is loose. Relationships remain undefined, a lack of clarity that is as much a measure of ecological malaise as the present state of the grizzly bear. But there are three certainties to guidc us toward home. First, ecosystem management will not be supported by a system rooted in resourcism, endless material and population growth, and political inequality. Dr. David Johns, a professor and activist living in Portland, Oregon, offers these words: "We need to grasp the necessity of moving beyond the choices offered by the powers that be; only by pushing beyond the limits of what is acceptable to the existing political-economic order can constraints on ecological-political choices be transcended."[95] This cannot happen without broad-based public participation and support. Second, though the demands of ecosystem management may seem overwhelming, hope is kindled everywhere by example: People from the Greater North Cascades to Zambia are rising up even now to resist the loss of ancient forests, prairies, zebras, and elephants. There has never been such a swelling of grassroots activism as exists today. The third fact, final and inescapable, is also the most reassuring. The biodiversity crisis has given us a lucid definition of who "the people" are; we are inextricably joined with the rest of life on Earth. The connection is more than momentary. For even as the biodiversity crisis provides the insights with which to create ecosystem management today, it points beyond the present toward what kind of people we may become tomorrow.

CHAPTER SEVEN

RESOURCES, ECOSYSTEMS, PLACE

Sometimes I go about in pity for myself,
and all the while a great wind is bearing me
across the sky.
Ojibwa song

It was by pale morning light during the time of the Great Flood that the First People encountered Koma Kulshan, the Great White Watcher, gleaming in the east. After drifting with the tides for weeks the First People followed the beaconlike cone of the mountain to safe harbor in a new land. Today, we call the peak that watches over the northwest corner of the Greater North Cascades ecosystem Mount Baker, in honor of a lieutenant in Captain George Vancouver's 1792 exploration party.

The roots of relationship between people and the land are found in the contrast between the mountain as sacred and the mountain as profane. According to the one view, Koma Kulshan is vital to human's identity-in-place. Without its guidance the First People would not have found a home to dwell in. The mountain patterned their lives. The story of its guiding power, handed down through generations, sustained a relationship that now for the most part is forgotten. According to the other view, the modern one, Mount Baker is a backdrop to our busy lives. It is so much scenery from the interstate or ferry, a convenient year-round recreation area accessible by automobile.

How close can we get to Koma Kulshan today? Is the Great White Watcher accessible to us, is he still alive? What is our proper relationship with the other mountains, the ancient forests, and the nonhuman beings of the Greater North Cascades? Asking these questions and attempting to come up with vital answers to them reveals the depth of the biodiversity crisis. These are not matters of abstract philosophy, idealism, or wishful thinking. They cannot be resolved by simply restructuring land management. To know Koma Kulshan, to experience the Greater North Cascades as home, one must be ready to change one's life.

The Council of All Beings

> *grey wolf*
> *howler monkey*
> *sperm whale*
> *blue whale.*
>
> *Give me a story . . . a song for a sadness too vast for my heart, for a rage too wild for my throat.*[1]

A cold wind rolls off the ridgeline and wraps the lodgepole forest in a chilly shawl. The moan of a single, measured drumbeat rises from an opening in the thin woods and mingles with the breeze, pulsing slow as death.

> *grizzly bear*
> *Florida panther*
> *swallowtail butterfly*
> *Utah prairie-dog.*

Hide me in a hedgerow, badger. Can't you find one? Dig me a tunnel through leaf mold and roots . . . My heart is bulldozed and plowed over.

One hundred people sit in a rough circle, grieving lineages lost beyond recall. Cries of anguish pierce the pine bark, leaden the sky. The drum chants the roster of extinction and the list goes on and on.

hawksbill turtle
Nile crocodile
red kangaroo

The pain cuts like a dull knife. Some people cry out "No!" Others shake from between the shoulders, bury their heads in their hands. I smell fresh grass, earth, and sorrow beyond words.

fan-tailed flycatcher
ferret
curlew
cougar
bobcat.
Your tracks are grown fainter. Wait, wait . . . Don't leave us alone in a world we have wrecked.

This is our council, prayer, lament. It is an act of defiant despair for all the ghost bears, birds, and butterflies now gone from our midst.

The rain stops. The chanting ends. The drum ceases. We are bid to come together, to embrace our pain and that of our councilmates, to hold each other through this grief that, faced, will make strong our defense of those beings that still remain alive.

A ragged procession snakes its way down the dead-end logging road, away from Goodell Creek, following the beat of the drum to another council. Morning sunlight filters through dense second growth. Illuminated sword fern and salmonberry lie scattered across the woods. A blue sky promises afternoon heat. The group shuffles silently through light and shadow. I see Bear, Northern Flying Squirrel, Red Cedar, Stetattle Creek, Salal, Bracket Fungus, and others. They struggle to keep the drum's rhythm. The drum leaves the road, guiding the group through a screen of second-growth alder to a grassy clearing shielded by young trees. In the open light each being becomes half human: Bracket Fungus is Susan, Squirrel is George, Grizzly Bear Rachel.

We sit in a circle, animals speaking with human voices, responding to the incessant beat of the drum, centuries of lost habitat, brethren gone down before guns, grazing, and human expansion. Our pain is buried deep. It becomes denial, muting the voice as sure as death. Furtive glances say, This is not my loss; the drum is inauthentic, a sham; I am not red cedar; Stetattle Creek feels no pain; this mask is fake; my life is separate from these others. This small group of people, bound together during a weekend retreat, is unable to express its collective connection with those very species that give reason to the struggle to protect the Greater North Cascades. Denial replaces despair; the animals drop their masks and become human again. The council is scuttled, the circle broken. Salal and Cedar slip off into the woods. Rachel, George, and the rest of the group stand up and listlessly follow the drum back toward camp. There is no moss-lined trail through pine woods here, only the potholed gravel logging road leading past cutover forest.

Ecosystem Management for the Ecological Self

First conceived and practiced in 1985 in Australia, and led by deep-ecology activists John Seed and Joanna Macy, the Council of All Beings is a ceremony of death, acceptance, and rebirth to action.[2] It begins with participants bearing witness to the biodiversity crisis—the extinction of species and destruction of wild habitat. They confront these losses, acknowledge their despair, and together exorcise their pain. In the words of the Buddhist monk Thich Nhat Han, the ritual allows people to "hear within themselves the sounds of the earth crying." Pain and loss are shed as the celebrants cultivate their link with species from the local ecosystem where the council takes place.

At Goodell Creek, after spending a few hours alone, each member made a mask that represented a species he or she had discovered in some special way. Later, seated in a circle, the participants were to speak for their chosen species by relaying messages from the wild to the world of humans.

Not every council is successful. Communications were heartfelt and clear at the Lodgepole Council, but they were garbled and unintelligible at the Goodell Creek gathering. People there were unable to enter into the spirit the ceremony, to unburden themselves of the need to remain rational and in control. At Goodell, the recitation of lost and doomed species pushed people in the direction of mental models and conceptual explanations. They felt manipulated. Grizzly Bear disappeared.

The Lodgepole Council changed my life. Goodell Creek stopped me in my tracks, and reminded me that, while the biodiversity crisis deals with ghost species and damaged lands, it is also about people who have little conception of what membership in a natural ecosystem means.

After decades of urbanization, Americans today lack the most rudimentary understanding of ecology. For seven years I have given a quiz on basic natural history to undergraduates at the beginning of field ecology courses to determine the level of their knowledge. Most cannot answer questions like What are the plant communities in your neighborhood? or Where does your water come from? Ecological illiteracy, combined with ignorance of the American political process, yields an almost complete lack of awareness of the purpose of parks and reserves, much less how they are managed. Over the last several years I have quizzed students majoring in environmental studies about the plans for their favorite natural area—98 percent have no idea what is in store for the future. The majority do not recognize the difference between the Forest Service and the Park Service. Land managers themselves are trained in natural resource and forestry schools, where the emphasis is on ecosystems as commodity producers. Only a few U.S. forestry schools offer conservation-biology course work. And the bureaucratic environment of professional land management only reinforces the notion of human domination of nature. These trends have profound implications; there is literally no *place* for bears and wolves to live in a culture that does not value them deeply.

The biodiversity crisis may change all of this. Ecology and conservation biology, so far, have provided compelling evidence that "we are enfolded . . . within the living, terrestrial environment."[3] Because this conclusion is founded on scientific theory and research, it is that much more difficult to deny. For years the Earth First! movement proposed big-wilderness protection and restoration based on deep ecology principles and the inherent rights of nature.[4] These visionary ideas were ignored by managers and policymakers and branded as utopian. Now, some conservation biologists are beginning to call for national ecosystem protection policies that render even the Earth First! proposals inadequate.

Members of the deep-ecology movement propose that ecological understanding and advocacy can contribute to a more engaged relationship with nature.[5] Humans have an innate capacity to identify with a wide range of other beings: blood relations, friends, the family of Homo sapiens worldwide. Because of anthropocentrism, this capacity stretches little beyond social identity in modern mass culture. Deep-ecology supporters seek to expand the social self in the direction of an ecological self. Who is that?

Think of someone you have loved for a long time. At the beginning of your relationship there was simple recognition of physical attributes and other basic character traits. Later, you began to relate to your friend. As your friendship deepened over time, you may have come to identify with that person. This process of growth, familiar to all, is not limited to our own species. When it expands beyond human intercourse to embrace relationships with other beings, the ecological self is born. Silas Goldean says, "Identity is not an airy generalization; it is grounded in the specifics of daily living, the solid particulars, blood and soil and sun. The only way to grasp this is to live somewhere long enough to attend to the complexity of living relationships."[6] The poet Robinson Jeffers called this "falling in love outward."

Deep ecology suggests that what makes us possible—air, soil, sun, water, other beings—must be part of ourselves. We must learn to recognize our ecological selves. In the Greater North Cascades ecosystem, old-growth forests nourish northern spotted owls *and* human beings. Yet clearcutting on the scale that is practiced today is a knife to the throat of the ecological self; soil washing off steep slopes is a part of the larger self lost forever. Reciprocity is implicit in the laws of ecology: Extending our identity increases our responsibility to *all* parts of the ecological self. Then, as philosopher Arne Naess says, we "may come to see [our] interest served by conservation, through genuine self-love, love of a widened and deepened self."[7] This is the underlying possibility of the biodiversity crisis: to "cultivate the affirmation of life and to sustain it in one's heart for a lifetime."[8]

The biodiversity crisis will be revolutionary only to the extent that humans discover the ecological self and replace adversarial relationships toward the Other with reciprocity. As philosopher Thomas Birch points out, the Other does not have to be an adversary. Other peoples

> see the land in terms of a different story, a story which holds that the fundamental human relationship with nature . . . is participatory, cooperative, and complementary, rather than conflictive. At times, of course . . . there is conflict, but normally wild nature sustains, sponsors, empowers, and makes human existence possible. Nature is wild, always wild, in the sense that it is not subject to human control. In this sense, humans are participants in a wildness that is far larger and more powerful than they can ever be, and to which human law-bringing is so radically inappropriate as to be simply absurd.[9]

The result of adversarial law bringing—our current land-management practices—is a last handful of grizzly bears in the Greater North Cascades, the imminent loss of thousands of North American plants and animals, endangered ecosystems, and at the planetary level, catastrophic climate change that may be irreversible over lifetimes. This is the ecological reality of the biodiversity crisis.

Planetarywide environmental destruction is mirrored by my students' ignorance of the birds in their backyards and of the future of the few wild places that remain. In the 1940s Aldo Leopold could comment that "to be an ecologist was to live in a world of wounds," but even he could not foresee what living in urbanized and degraded landscapes would do to humans. Today, writer Barry Lopez notes that "year by year, the number of people with first-hand experience in the land dwindles . . . herald[ing] a society in which it is no longer necessary to know where they live except as those places are described and fixed by numbers."[10] People are losing the ability to relate with nature tactilely and spiritually.

From the standpoint of healing nature and cultivating the ecological self, land-management reform will fail if it merely substitutes scientific ecosystem management for resource conservation. There is a great difference between creating biodiversity-protection networks and recognizing greater ecosystems following biotic boundaries. Protecting the wildlands surrounding Mount Baker is not the same as experiencing Koma Kulshan. Implementing ecosystem management for native diversity, difficult as that promises to be, is a short-term goal, the initial step of a protracted process that may last longer than the time it takes to restore sufficient ancient-forest habitat for the northern spotted owl. Defenders of biodiversity need to cultivate a long-term view and see biodiversity-protection networks and endangered ecosystem laws as restorative acts for humans and nonhumans alike.

In the short term, biodiversity will be further reduced as the pace of extinction quickens and human population growth and resource consumption continue unabated. It is a sad irony that we seem only to comprehend the self-destruction implicit in our values as the world unravels around us. There is no ultimate solution to human ignorance, nor to the practical problem of how to live sustainably. Education and developing ecological awareness are ongoing, open-ended processes—there is no end to enlightenment. This is part of the paradox of being human. Some deep ecologists believe that once "people attain a deep-seated realization of [the ecological self] they will scarcely be able to refrain from identifying with 'other' entities,"[11] but there is nothing utopian or final about living as if nature mattered. There are only choices and more choices to be made in a diverse and uncertain world.

The Fallacy of Ecosystem Management

The grizzly bear in the Greater North Cascades, once strong and numerous enough to be respected as a teacher by the First People, the

Methows, the Okanogans, and other eastside tribes, is now a spirit without form. No one has definitely seen one since 1969. Some say that this is as it should be, that progress has come to the Pacific Northwest and that the Bear Mother myth is anachronistic at best in a world of computers, video games, and transcontinental jet travel. There is nothing for modern people to learn from old stories.

These same critics side with Hal Salwasser, Jim Agee, and Darryll Johnson, and the rest of the process-ecosystem-management theorists when they put trust in the power of the emerging ecosystems view of nature to stop the erosion of biodiversity.[12] As we rush toward the twenty-first century, this solution is simple: increase funding for conservation-biology research, hire more staff ecologists, draw new boundaries and convene interagency committees. But there is little to glean from these acts alone. The scientific ecosystems view, which grows directly out of our old environmental values, is pre-adapted to management-as-usual. Grizzlies and greater ecosystems need more than land-management reformation.

For the decade of the 1990s, the choice is clear. Either we begin to embrace the revolutionary aspects of protecting biodiversity—the challenge of saying "less" to population growth and habitat destruction, of questioning the ethic of domination—and seek education for the ecological self, or we create almost by default a world where one animal (the human being), having appropriated to itself the fate of other species and ecological processes, causes the degradation of all. There can only be one result: overmanaged ecosystems, natural zoos.

This is the "lab mentality" of managers stretching all the way back to Gifford Pinchot. It operates when people see lumber for forests, ore for rocks, real estate for landscapes, labor for people, and "well-having for well-being."[13] It is strengthened when managers blame the loss of species on poor organization and inefficient bureaucracies. It is upheld when biologists confuse the map (data, models, and GISs) with the territory ("the real, the immediate, and the mythic dimensions of species and places").[14] Couple it with the managerial demands of averting the extinction of tens of thousands of species or reversing global warming and the entire world becomes a laboratory within which humans can refine the art of controlling nature.

We must be very clear in our thinking about management. It is encouraging to see an ecosystems view being debated. But it is still too soon to tell what set of values ecosystem management will ultimately support. For, even though scientific ecosystem management may flourish in the coming years, biological diversity will not be sustained if new ways of managing nature do not also transform how we experience our place in nature, how we manage ourselves. Here are the reasons why.

The need for more scientific understanding of key issues in conservation biology is obvious. For the near term, we cannot hope to counter threats to biodiversity without more information resulting from increased research. But our environmental values drive us to adopt information bereft of wisdom. Even as we lack the ability (and funding) to solve difficult theoretical dilemmas, we have not broached the question Where does this path lead us? Our belief that the world is here primarily for our use does not require such speculation. Yet, the sheer amount of research and the concomitant scale of management needed to "solve" the biodiversity crisis scientifically go far beyond the familiar bounds of cost and benefit. At some point, the level of knowledge and manipulation required in a human-dominated world runs into a contextual paradox: No single part can control the system within which it is embedded.[15]

Holmes Rolston suggests that "an ecologically-informed society must love lions-in-jungles, organisms-in-ecosystems or else fail in vision and courage."[16] The sort of ecosystem management that plans to radio-collar every grizzly and track every wolf dooms our vertebrate kin to a piecemeal existence. The trend is incremental and therefore insidious: The more we confine ourselves to fragments, the less we are able to fathom wholes. We fail to see that "laboratory-manipulated artifacts are not ecologically self-sustaining beings."[17]

What will be left wild if we manipulate everything? If the main goal of ecosystem management is the protection of biodiversity, we need to look far enough into the future to see the implications of decisions made today. Grizzly cubs, human children, salmon fry, and wolf pups all have vital needs that bind their future with today's land management practices. Yet our assumptions and the crisis mentality of the present seem to "require" us to support captive breeding strategies and politically driven wolf-reintroduction plans.

Many managers would agree with conservation biologist Michael Soulé that we should use whatever means necessary to "save as many pieces of the planet as possible and argue later about whether saving species by artificial means was in their best interests."[18] Soulé predicts an *ex situ* "cryoconservation" future where, owing to advances in biotechnology, "entire communities . . . might be put in a zip-lock freezer bag." The founder of the Society for Conservation Biology reasons that genetic engineering will continue to achieve rapid breakthroughs and that the "term 'natural' will disappear from our working vocabulary." He believes that the word is "already meaningless" because no part of the biosphere has been left undisturbed by human activity.

Is a high-tech, genetic, *ex situ* future inevitable? Is the word *natural* meaningless? Are all strategies equally appropriate in the struggle to protect diversity? Soulé's provocative statements spotlight the complexity of the debate over ending the biodiversity crisis. It also illustrates how easy it would be for us to use scientific ecosystem management to increase our domination of nature.

My view differs from Soulé's. I don't believe that scientific understanding of nature will continue to grow by leaps and bounds. Soulé expects present levels of research funding to continue to finance high-tech science. But in the United States the staggering budget deficit, public education, toxic-waste cleanup, and a decaying industrial and transportation infrastructure, to name but a few, are competing for scarce funds. Some economists estimate that the cost of limiting atmospheric carbon buildup alone will run into the *trillions* of dollars through the next century.[19]

The word *natural*, while not easy to define, is far from meaningless. What is natural depends not merely on the degree to which people have affected ecosystems in the past or may alter them in the future.[20] Nature will always be primordial, spontaneous, generative—there is nothing on Earth that is unnatural. Yet we can distinguish between what is wild and what has been tamed, between spontaneity and design. Human activity changes the course of spontaneous nature every day. In this sense, the Greater North Cascades ecosystem is an artifact, the result of a historical relationship between people and nature in the Pacific Northwest. When we forget this, we fall prey to the preservationist illusion that there is a pristine nature that exists apart from Homo sapiens. Studies of the distribution and abundance of mountain lions and mule deer in the Great Basin of Nevada show how management for pristine conditions can be an elusive endeavor.[21] Nonnative cattle, sheep, and horses contributed to overgrazing and ecosystem degradation, and as a result the native populations of lions and deer increased. Environmentalists pressure the government to eliminate introduced species, but no one argues for a reduction in native species that breed more successfully as a result of human activity.

The meaning of *natural* is also relative. Biologist Jay Anderson reminds us that there is a continuum between complete naturalness and complete unnaturalness.[22] Organic farming, in contrast to farming with pesticides and fertilizers, is more in tune with local nutrient cycles and waste streams. Yet both types of farming depend to a greater or lesser extent on fossil fuels: "natural" is a relative condition.

Some strategies to solve the biodiversity crisis are better than others. Our future will be dominated by biotechnology if we forget that scientific solutions to the biodiversity crisis, whether through freezer-pack preservation or big wilderness reserves, are strategic, not ultimate. Science can census northern spotted owls and recommend habitat-conservation areas, but it cannot make final choices—that is a job for humans, who value one set of outcomes over another. The biodiversity crisis is pushing us to decide whether to continue our attempt to tame nature or to change course, pull back, and become a partner with nature.

In periods of crisis, when time is short, the stakes are high, and social consensus is lacking, decision making is exceedingly difficult. Settlements become protracted, conclusions are postponed. The problem is that we learn and practice ecological lessons at a rate slower than the rate of our population and consumption increase. For this reason, we will not have the luxury of arguing the merits of the natural zoo versus native diversity at some point in the future. For many endangered species and ecosystems there is no later—the time for action is now. In some cases, a combination of *in situ* habitat protection and *ex situ* high technology may benefit a species. Still, we have to keep our efforts in perspective. We have worked to save the California condor through an expensive captive-breeding experiment that includes some new habitat acquisition. But even as the first birds were released into the wild in 1992, there was no plan to ensure enough habitat for a self-sustaining viable population in the future.

Restrict the definition of *natural* to degrees of anthropogenic disturbance, seek captive breeding for all keystone species, freeze-dry ancient forests, and that is the world future generations will inherit, experience, and value. Aldo Leopold recognized this when he noted, "We can be involved only in relation to something we can see, feel, understand, love, or otherwise have faith in."[23] The Japanese poet Nanao Sakaki underlines this point in a poem:

> At a department store in Kyoto
> One of my friends bought a beetle
> For his son, seven years old.
> A few hours later
> The boy brought his dead bug
> To a hardware store, asking
> "Change battery please."[24]

Give a boy a beetle in a box. Remove the grizzly from the Greater North Cascades. Capture the last condor. Clone a full complement of blue-whale genes and reduce the Pacific Ocean to a swimming pool. For humans, one species among many, the limits of life disappear.

Dwelling in the Commons: A Map for the Future

I have heard it said that you can use nature, like the Bible, to prove anything. There is no need, however, to prove that 30 million American bison no longer roam the Midwest, that old-growth Douglas-fir forests are an endangered ecosystem, or that the continental U.S. *Ursus arctos* population is down to its last 1 percent. All are almost gone. Ecological lacunae are obvious, though the overall effect on human health and well-being may be more difficult to detect. The biodiversity crisis offers an opportunity to measure ecological degradation as well as to monitor people's ability to perceive the consequences. The threat of scientific ecosystem management is that it will preempt the possibility of learning to live sustainably. Ecosystem management is not about better control of elk populations or more efficient fire management. It is about acquiring the wisdom to understand the limits of human endeavor. For centuries we have invested in the anthropocentric attempt to escape the limits of our home, Earth, regardless of the costs to other beings. We have changed the definition of *sustainable*: While it once meant "to keep in being," it now means managing "all assets for increasing [human] wealth and well-being."[25]

The biodiversity crisis, though it may be experienced less as a bang and more as a gradual winking out of wildness, is likely our last reminder of the need to "learn self-discipline and caution in the face of desire and availability."[26] The choice is ours—a planet where the gap between people and nature grows to a numbing, incomprehensible chasm, or a world of damaged but recoverable native diversity where the operative word is hope.

Ecosystem management should not be the end product of our inability to control ourselves. It should be the first step toward discovering place, coming home. It should be a call to restorative action on behalf of genes, grizzlies, and greater ecosystems, a call to companionship with the ecological self, an invitation to take on responsibility, literally to become able to respond to those others that we have for so long denied.

What is place? Where is it? If you meet the Great Bear on equal terms, what do you do? How do you learn from the northern spotted owl? Douglas-fir?

An ecologist might define human's place as living within the constraints of local and regional ecosystems to satisfy basic needs. A bioregionalist might identify the experience of place as "becoming native . . . through . . . applying for membership in a biotic community and ceasing to be its exploiter."[27] A Mexican grassroots activist would say that place is regeneration by substituting trust and friendship for "rules of access" and the myth of progress.[28] A Gitksan Indian woman from British Columbia might suggest that a sense of place goes hand in hand with responsible decision making that involves all beings, not just humans.

Throughout this book I have attempted to describe a path away from nature-as-commodity toward ecosystem management for native diversity. This step will challenge us to live up to our epithet, Homo sapiens, "wise animal." The shift from a scientific ecosystem perspective to one where the split between people and nature is more fully healed, where humans develop a true sense of place, will be more formidable still. This is structural change, change at the level of belief and values or, more to the point, of our direct experience of being in the world. Environmental philosophers call it ecocentrism when a person experiences life as a member of an extended ecological community rooted in a specific place instead of as one individual competing against many others.[29] I cannot provide the details of how humans will make the transition from ecosystem to place. As the Gitksan woman would tell it, the particulars of place are identified in living, and we have much to discover.

How do we begin? Here is a map and a story:

The world is an infinitude of specific places. In the Greater North Cascades ecosystem, *this* gravel bar spawns Canyon Creek cutthroat trout. In summer, *this* fir branch bears Townsends warblers. One-thousand-year-old cedars inhabit *this* lower tongue of Noisey Creek. Columbia River melts through *this* lava-rimmed watershed. There are rubies up Granite Creek, giant trees to log across the entire westside, and, possibly, a grizzly bear or two living north of Mount Baker. There are also certain Skagit Valley raspberry fields, waterfront docks in Bellingham, a growing Puget Sound supercity, and two sacred mountains, Koma Kulshan and Tahoma (Mount Rainier).

In the days before the great split between people and nature, each human dwelling had a central hearth.[30] The world was experienced from the fire pit outward—yard, neighborhood, region, lands in joint use, and, finally, the wild, the place "where the bears are." The human population was low enough to allow access to all these parts, from the domestic fire to the shared berry patches and wood lots of the commons to backcountry solitude.

Today, this world has been supplanted by a less diverse landscape shaped by law and ownership and streamlined for efficient economic production. Our sense of place has atrophied. The hearth and the neighborhood have become private land, the commons has been replaced in the United States by federal ownership. The backcountry is now "roadless area," a small subset of the public domain, sometimes enclosed legally as wilderness, sometimes ripe for development. In England, from the fifteenth to the mid-nineteenth centuries, the feudal aristocracy with the support of the crown captured the commons through a series of enclosure acts that forced the country people toward the cities and consolidated ownership for industrial production. The U.S. frontier was declared closed in 1890. Soon after, the Forest Service, the Park Service, and centralized land management were born.

Federal administration does not resemble the sort of management practices that preceded national forests and parks. Anthropological research worldwide has shown that there existed many varieties of commons management prior to the rise of commodity-oriented societies.[31] The two outstanding features of most successful, long-term cultures that worked with nature were intimate knowledge of plants, animals, and ecosystems and small-scale community control. Awareness of ecological specifics helped people experience the *where* of bears, herbs, and fuelwood, the *why* of soil erosion and hunting success, and the *when* of enough, the limits of human intervention. Decentralized regulation allowed direct knowledge of the effects of management and promoted widespread understanding, acceptance, and enforcement of community rules of use.

Enclosure replaced this sort of ecosystem management with a new set of standards. We have been taught that the public lands are always open, that there are few social restrictions on their use, that users are selfish and always seek to maximize short-term gain, and that under this condition exploitation is bound to occur. This view, and the shift toward urban experience and contractual obligations that came with it, has hidden the commons from us. With the land enclosed, the old institutions destroyed, and population growth and resource consumption on the rise, it is no wonder that people today take these assumptions about management as the truth and pattern their lives accordingly. The original map of the commons has been redrawn—legal and political rules of access have replaced patterns of community-controlled land use and place-based social relations.

This modern map is superimposed on the land. But places remain alive, salmon still swim upstream, and this coming spring an owlet will find a new home inside an old-growth snag. Ecosystem management, if we aim to protect species and ecosystems, will begin to fashion a new map of nature when the buffer zones that must be part of a biodiversity-protection network begin to feed the system *both* ways—toward the wildlands at the core of the network and the modern heartlands of towns and cities. The lesson here is that the ecological and social selves are interdependent. Both have atrophied—we have lost the commons along with our very sense of human neighborliness.

If we begin to gain direct experience with ecosystems by working to protect biodiversity, we may find that nature becomes part of culture again. There is little reason to restrict the knowledge of plants and their uses to botanists and ecologists. The crucial difference between the old culture of the commons and modern industrial civilization lies in what anthropologist Richard Nelson discovered while working with the Koyukon people of northwestern interior Alaska. Nelson noted that the Western definition of wilderness reflects our "inability to conceive of occupying and utilizing an environment without fundamentally altering its natural state. But, the Koyukon people and their ancestors have done precisely this over a protracted period of time . . . they have made a highly effective adjustment to living as members of an ecosystem, pursuing a form of adaptation that fosters the successful coexistence of humanity and nature within a single community."[32] This glimpse of the past is a glimpse of the future. When humans are able to adapt to the conditions that limit existence, when once again the commons is held by all creatures, then we will replace scientific ecosystem management with place.

I was walking in early dawn at the edge of the world where the upthrust cliff line of Mesa Verde plunges to the Montezuma Creek plain. The sun threw a blanket of rose-tinged light over the farms and irrigated fields that spread toward the broken arroyos at the base of the mesa. Another day in the land above the desert where winter snow and summer rain grow oaks and pines out of the rocks.

The Park Service interpretive trail I was following took me away from camp, out through a line of sandstone cliffs that rose up sheer to the north, and along the western edge of the mesa, winding in and out of scrubby thickets, boulders, and bunchgrass. Ute Mountain shot its dark volcanic body into the sky. The south wind blew sagebrush into my nostrils. Passing the cliffs, I stopped to admire their steepness, wonder if they harbored ravens' nests, puzzle out their rocks—were they Mesa Verde, Coconino, or Cliff House sandstone? The rocks did not reveal themselves, and I couldn't find a Park Service post that keyed into the trailhead booklet describing the geology and natural history of the Knife Edge trail. I walked on. At the end of the trail, after observing two brown towhees foraging for breakfast—the mesa shadow was giving way to morning light—I headed back.

Where the Knife Edge cuts through the cliffs I met Loren standing in the trail, watching the rocks. We were friends camping together, both first-time visitors to Mesa Verde National Park. Loren is a Pomo Indian *yomta*, a dreamer and healer from Kashaya, near the Sonoma Coast of California.

"Hey, did you see those cliffs over there?" He pointed to the nameless sandstone.

"Sure did," I replied. "I don't know what they are but they're hard to miss. I wonder why they're there?"

Loren is a humble man. He carries the responsibility of spiritual leader of the Kashaya Pomo with quiet dignity. "Well, Ed, I was stopping here to ask these cliffs permission to walk past. They have the energy of this place and I wanted to make sure to stop and ask them if it was okay to go by."

I looked at the cliffs, their color canvas-tan. Old ocean-floor sediments from millions of years ago right here in this place. If I could stand in this spot for centuries I would witness boulders tumbling down, generations of ravens and hawks, cliff rose, and yucca plants, beating wings and probing roots. The cliffs would remain.

"Yes," I said, nodding. Loren smiled, looked past me as if he half expected me to laugh or walk away.

"Tomorrow," he said, "I'm supposed to bring a group from camp out here and tell them about what I do, about this energy I feel. I thought it would be good to walk here this morning and see what was going on.

These rocks, they'll help me with those people tomorrow, I think. You've been here before. I thought I'd check with you. What's down this trail?"

I told him about the gambel oak thickets, the rose hips, the late-flowering rabbit brush, and the towhees. The morning sun was warming the boulders around us.

"I'll see you back at camp. Don't eat all the oatmeal," he laughed.

Loren walked out on the Knife Edge. I headed back for breakfast. The cliffs stood still. There was no Park Service interpretive sign.

The Promise of the Biodiversity Crisis

How do people change?

Sandstone, grizzly bears, greater ecosystems, elders—the teachers are everywhere. I do not pretend to believe that place-based cultures will reinhabit North America in the near future. But the tree of life is being shaken. The power of the biodiversity crisis is in providing non-negotiable evidence against human domination of Earth. Deep ecology theorizing and the rise of environmental ethics in academic philosophy are encouraging trends, although they are not as compelling as massive clearcuts in northern-spotted-owl habitat, toxic waste destroying a community aquifer, or the poaching of African elephants. So far, scientific ecosystem management has responded to these threats with the abstract language of reform. The legacy of multiple-use does not permit immediate implementation of participatory ecosystem management any more than a call for zero population growth reduces human numbers overnight. Land management bureaucracies, mainstream environmental groups, a slow-moving Congress, and an uneducated citizenry will not disappear tomorrow. Yet the promise of the biodiversity crisis remains, for as philosopher Michael Zimmerman reminds us, "There is no contradiction in seeking to reform current ways of using nature while simultaneously seeking a shift towards a more inclusive level of awareness, in light of which nature would manifest itself as something other than an object for domination."[33] Nor is there anything utopian or novel about "pursuing a form of adaptation that fosters successful coexistence" between people and ecosystems. Some humans have been doing this for thousands of years. It is simply necessary.

What *is* necessary? In one place, the Greater North Cascades, where rivers tumble down from ice and ancient forests stretch toward the stars, this single query breeds a diversity of questions: How many grizzly bears live in the mountains, and what are their vital needs? What are the boundaries of the Greater North Cascades ecosystem? What is the proper role of humans in this bioregion? What are the optimal populations of *all* native species? How might sustainability be built into present ecosystem management practices? How can we identify with the ecological self?

Outside of these questions lies Koma Kulshan, the heart of wild nature, with lessons that go beyond maps, stories, and the simple prescriptions of management plans. Thoreau lamented that "we live here to have intercourse with rivers, mountains, beasts and people. How few do we see conversing with these things." Even as we gather to decide the immediate fate of the Greater North Cascades, even as we bear witness to the many threats to native diversity and offer plans in its defense, the mountain stands, waiting.

Sometime in the future we won't simply get in a car, drive to a trailhead, go backpacking, and return to a city. Not long from now, deep in an old-growth forest, we will come to experience what the poet Mary Oliver felt—that "whatever my place in this garden, it is not to be what I have always been—the gardener."[34] In some future time there won't be any national forests, national parks, or even a Greater North Cascades ecosystem. There will simply be home.

NOTES

Chapter 2: The Biology of Thinking like a Mountain

1. H. Salwasser, "Editorial," *Conservation Biology* 1 (1987), 275.
2. D. Chadwick, *The Biodiversity Challenge* (Washington, D.C., 1990), 5.
3. N. Keyfitz, "The Growing Human Population," *Scientific American* 261: 3 (1989), 119. See also P. Ehrlich, *The Population Explosion* (New York, 1990).
4. M. Soulé et al., "The Millennium Ark: How Long a Voyage, How Many Staterooms, How Many Passengers?" *Zoo Biology* 5 (1986), 102.
5. N. Johnson et al., "Alternatives for Management of Late-Successional Forests of the Pacific Northwest." Report to the Agriculture Committee and the Merchant Marine and Fisheries Committee of the U.S. House of Representatives, 102d Cong., 2d sess., October 8, 1991.
6. P. Ehrlich and E. O. Wilson, "Biodiversity Studies: Science and Policy," *Science* 253: 5021 (1991), 758–62.
7. R. L. Brown and C. Wolf, *Soil Erosion: Quiet Crisis in the World Economy* (Washington, D.C., 1984). Paper 60, Worldwatch Institute.
8. E. Norse, *Ancient Forests of the Pacific Northwest* (Washington, D.C., 1990), 272.
9. N. Myers, "A Major Extinction Spasm: Predictable and Inevitable?" in D. Western and M. Pearl, eds., *Conservation for the Twenty-first Century* (New York, 1989), 42–49.
10. J. Diamond, "Overview of Recent Extinctions," in Western and Pearl, *Conservation for the Twenty-first Century,* 38.
11. D. Raup and J. Sepkoski, "Mass Extinctions in the Marine Fossil Record," *Science* 215 (1982), 1501–3.
12. D. Simberloff, "Are We on the Verge of a Mass Extinction in Tropical Rain Forests?" in D. Elliott, ed., *Dynamics of Extinction* (New York, 1986), 165–80.
13. M. Soulé, "The Millennium Ark," 102–4.
14. E. Norse, *Ancient Forests,* 257. Norse uses the figure 87 percent here, but I updated it to reflect cutting since 1989.
15. J. Robbins, "Pollution Shrouding National Parks," *New York Times,* December 3, 1989, 19.
16. R. Graham et al., "How Increasing CO_2 and Climate Change Forests," *BioScience* 40 (1990), 575–87. For an overview, see R. Peters and T. Lovejoy, eds., *Global Warming and Biological Diversity* (New Haven, 1992).
17. M. Davis, "Insights from Paleoecology on Global Change," *Bulletin of the Ecological Society of America* 70 (1989), 222–28.

18. W. Westman, "Managing for Biodiversity," *BioScience* 40: 1 (1990), 30.
19. Office of Technology Assessment (OTA), *Technologies to Maintain Biological Diversity* (Washington, D.C., 1987).
20. J. Franklin et al., *Ecological Characteristics of Old Growth Douglas-fir Forests* (Portland, 1981), U.S. Forest Service general technical report PNW–118.
21. R. Noss, "From Endangered Species to Biodiversity," in K. Kohm, ed., *Balancing on the Brink of Extinction* (Washington, D.C., 1990), 227–47.
22. See in general T. Allen and T. Starr, *Hierarchy: Perspectives for Ecological Complexity* (Chicago, 1982); R. O'Neill et al., *A Hierarchical Concept of Ecosystems* (Princeton, 1986); R. Noss, "Issues of Scale in Conservation Biology," in P. Fiedler and S. Jains, eds., *Conservation Biology: The Theory and Practice of Nature Conservation, Preservation, and Management* (New York, 1992), 240–50.
23. "Indicators for Monitoring Biodiversity: A Hierarchical Approach," *Conservation Biology* 4 (1990), 357.
24. A. Koestler, *Janus: A Summing Up* (London, 1978), 34.
25. D. Perry, "Bootstrapping in Ecosystems," *BioScience* 39 (1990), 230–37.
26. L. Harris, *The Fragmented Forest* (Chicago, 1984), 84.
27. M. Hemstrom and J. Franklin, "Fire and Other Disturbances of the Forests in Mount Rainier National Park," *Quaternary Research* 18 (1982), 323–51.
28. A. Leopold et al., "Wildlife Management in the National Parks," *American Forests* 69: 4 (1963), 32–35, 51–63.
29. R. Noss and L. Harris, "Nodes, Networks, and MUMS: Preserving Diversity at All Scales," *Environmental Management* 10 (1986), 301.
30. M. Soulé, "What Is Conservation Biology?" *BioScience* 35 (1985), 728.
31. R. MacArthur and E. O. Wilson, *The Theory of Island Biogeography* (Princeton, 1967). For the early history of conservation biology, see M. Soulé, "Introduction," in Soulé, ed., *Viable Populations for Conservation* (New York, 1987), and D. Simberloff, "The Contribution of Population and Community Biology to Conservation Science," *Annual Review of Ecological Systematics* 19 (1988), 473–75.
32. J. Diamond, "Island Biogeography and Conservation: Strategy and Limitations," *Science* 193 (1976), 1027.
33. Ibid.
34. W. Newmark, "A Land-Bridge Island Perspective on Mammalian Extinction in Western North American Parks," *Nature* 325 (1987), 430–32.
35. M. Shaffer, "Minimum Viable Populations: Coping with Uncertainty," in M. Soulé, ed., *Viable Populations for Conservation*, 69–70.
36. M. Gilpin and M. Soulé, "Minimum Viable Populations: Processes of Species Extinction," in Soulé and B. Wilcox, eds., *Conservation Biology: An Evolutionary-Ecological Perspective* (Sunderland, Massachusetts, 1980), 25.

37. M. Shaffer, "Minimum Viable Populations," 71; M. Shaffer, "Minimum Population Sizes for Species Conservation," *BioScience* 31 (1981), 131–34.
38. M. Soulé, "Introduction," in Soulé, *Viable Populations,* 1.
39. O. Frankel, "Genetic Conservation: Our Evolutionary Responsibility," *Genetics* 78 (1974), 53–65.
40. M. Soulé, "Thresholds for Survival: Maintaining Fitness and Evolutionary Potential," in Soulé and B. Wilcox, *Conservation Biology* (1980), 11–124; I. Franklin, "Evolutionary Change in Small Populations," in M. Soulé and B. Wilcox, *Conservation Biology,* 135–50; also see in general O. Frankel and M. Soulé, *Conservation and Evolution* (New York, 1981).
41. M. Soulé, "Thresholds for Survival," 111–24; I. Franklin, "Evolutionary Change," 135–50.
42. D. Simberloff, "The Contribution of Population and Community Biology to Conservation Science," *Annual Review of Ecology and Systematics* 19 (1988), 480–82.
43. B. Wilcox, "Extinction Models and Conservation," *Tree* 1: 2 (1986), 47.
44. D. Simberloff, "The Contribution of Population and Community Biology," 486.
45. D. Goodman, "The Demography of Chance Extinction," in M. Soulé, ed., *Viable Populations,* 11–34.
46. M. Shaffer, "Minimum Viable Populations," 73.
47. D. Simberloff, "The Contribution of Population and Community Biology," 497.
48. S. Jones, "The Implications of Hurricane Hugo on the Recovery of the Red-cockaded Woodpecker," *Endangered Species Update* 7: 1–2 (1989), 7.
49. W. Newmark, "Legal and Biotic Boundaries of Western North American National Parks: A Problem of Congruence," *Biological Conservation* 33 (1985), 197–208.
50. M. Soulé, "Where Do We Go from Here?" in Soulé, ed., *Viable Populations,* 179. For Soulé's current view, see Soulé and S. Mills, "Conservation Genetics and Conservation Biology: A Troubled Marriage," in O. Sandland et al., eds., *Conservation of Biodiversity for Sustainable Development* (in press).
51. R. Lande and G. Barrowclough, "Effective Population Size, Genetic Variation, and Their Use in Population Management," in M. Soulé, ed., *Viable Populations,* 87–123.
52. M. Soulé, "Introduction," 5.
53. M. Shaffer, "Minimal Population Size for Species Conservation," 133–34.
54. M. Shaffer and F. Sampson, "Population Size and Extinction: A Note on Determining Critical Population Sizes," *American Naturalist* 125 (1985), 150.

55. D. Mattson and M. Reid, "Conservation of the Yellowstone Grizzly Bear," *Conservation Biology* 5 (1991), 368.
56. D. Simberloff, "The Contribution of Population and Community Biology," 480. See also R. Levins, "Extinction," in M. Gerstenhaber, ed., *Some Mathematical Questions in Biology: Second Symposium on Mathematical Biology* (Providence, Rhode Island, 1970), 77–107; and M. Gilpin, "Spatial Structure and Population Vulnerability," in M. Soulé, ed., *Viable Populations,* 127.
57. M. Gilpin and M. Soulé, "Minimum Viable Populations," 19–34; M. Gilpin, "Population Viability Analysis," *Endangered Species Update* 6: 10 (1989), 15–18.
58. B. Wilcox, "Extinction Models," 47.
59. M. Shaffer, "Minimum Viable Populations," 70.
60. M. Soulé, "Introduction," 2.
61. L. Ginzburg et al., "Quasi-extinction Probabilities as a Measure of Impact on Population Growth," *Risk Analysis* 2 (1982), 171–81.
62. E. L. Braun, *Deciduous Forests of Eastern North America* (New York, 1950).
63. W. Newmark, "Legal and Biotic Boundaries," 197–208.
64. H. Salwasser et al., "The Role of Interagency Cooperation in Managing for Viable Populations," in M. Soulé, ed., *Viable Populations,* 159–82.
65. C. Schonewald-Cox, "Conclusions: Guidelines to Management, A Beginning Attempt," in C. Schonewald-Cox et al., eds., *Genetics and Conservation* (Menlo Park, California, 1983), 414–44.
66. Ibid.
67. G. Belovsky, "Extinction Models and Mammalian Persistence," in M. Soulé, ed., *Viable Populations,* 35–58.
68. M. Soulé, "Where Do We Go from Here?" 177.
69. J. Franklin et al., *Interim Definitions for Old-Growth Douglas-fir and Mixed Conifer Forests in the Pacific Northwest and California* (Portland, 1986), 4–5.
70. D. Wilcove, "Of Owls and Ancient Forests," in E. Norse, *Ancient Forests of the Pacific Northwest* (Washington, D.C., 1990), 77–78. See also J. Thomas et al., *A Conservation Strategy for the Northern Spotted Owl* (Portland, 1990).
71. D. Booth, "Estimating Prelogging of Old-Growth in the Pacific Northwest," *Journal of Forestry* 89: 10 (1991), 25–29. See also L. Harris, *The Fragmented Forest,* 29–30.
72. E. Norse, *Ancient Forests,* 244.
73. Cascade Holistic Economic Consultants, ed. "America's Forests: The Graphic Facts," *Forest Watch* 11: 5 (1990), 17.
74. U.S. Forest Service, *Land and Resource Management Plan: Mt. Baker–Snoqualmie National Forest* (Portland, 1990), plan glossary 26.

75. P. Morrison, *Old Growth in the Pacific Northwest* (Washington, D.C., 1988).
76. See G. Gray and A. Eng, "How Much Old-Growth is Left?," *American Forests,* September–October 1991, 46–48; P. Morrison et al., *Ancient Forests of the Pacific Northwest* (Washington, D.C., 1991), ii–iv.
77. B. Wilcox and D. Murphy, "Conservation Strategy: The Effects of Fragmentation on Extinction," *American Naturalist* 125 (1985), 884.
78. Ibid., 879–80; D. Wilcove et al., "Habitat Fragmentation in the Temperate Zone," in M. Soulé, ed., *Conservation Biology: The Science of Scarcity and Diversity* (New York, 1986), 238; J. Franklin and R. Formann, "Creating Landscape Patterns by Forest Cutting: Ecological Consequences and Principles," *Landscape Ecology* 1 (1987), 5–18.
79. R. Whitcomb et al., "Effects of Forest Fragmentation on Avifauna of Eastern Deciduous Forests," in R. Burgess and D. Sharpe, eds., *Forest Island Dynamics in Man-Dominated Landscapes* (New York, 1981), 125–205; J. Terborgh, *Where Have All the Birds Gone?* (Princeton, 1989).
80. See R. Yahner, "Change in Wildlife Communities near Edges," *Conservation Biology* 2 (1988), 333–39; M. Soulé et al., "Reconstructed Dynamics of Rapid Extinctions of Chaparral-Requiring Birds in Urban Habitat Islands," *Conservation Biology* 2 (1988), 75–92.
81. D. Wilcove, "Nest Predation in Forest Tracts and the Decline of Migratory Songbirds," *Ecology* 66 (1985), 1211–14. See K. Reese and J. Ratti, "Edge Effect: A Concept under Scrutiny," *Transactions of the 53rd North American Wildlife and Natural Resource Conference* (1988), 127–37.
82. See A. Hansen et al., "Conserving Biodiversity in Managed Forests," *BioScience* 41: 6 (1991).
83. T. Spies and S. Cline, "Coarse Woody Debris in Forests and Plantations of Coastal Oregon," in C. Maser et al., eds., *From the Forest to the Sea: A Story of Fallen Trees* (Portland, Oregon, 1988), U.S. Forest Service general technical report PNW-GTR-299, 5–23.
84. M. Raphael, "Long-Term Trends in Abundance of Amphibians, Reptiles, and Mammals in Douglas-Fir Forests of Northwestern California," in R. Szaro et al., eds., *Management of Amphibians, Reptiles, and Small Mammals in North America* (Fort Collins, Colorado, 1988), 12–31; M. Raphael et al., "Large-Scale Changes in Bird Populations of Douglas-Fir Forests of Northwestern California," *Bird Conservation* 3 (1988), 63–83.
85. U.S. Forest Service, *Land and Resource Management Plan,* iv–58.
86. P. Morrison, *Old Growth,* 24.
87. As quoted in P. Ford, "Holding On to the Seeds," *High Country News,* November 19, 1990, 24.
88. W. Ripple et al., "Measuring Forest Landscape Patterns in the Cascade Range, Oregon, U.S.A.," *Biological Conservation* 57: 91 (1991), 73–88.
89. R. Noss, "The National Forest Plans Versus Old Growth Protection:

Effects on Wildlife and Biodiversity." Unpublished report available from the Wilderness Society, Washington, D.C., 1991, 22–32.

90. M. Hemstrom and J. Franklin, "Fire and Other Disturbances of the Forests in Mount Rainier National Park," 32–51; P. Morrison and F. Samson, *Fire History and Pattern in a Cascade Range Landscape* (Portland, 1990), 2–3.
91. F. Bormann and G. Likens, *Pattern and Process in a Forested Ecosystem* (New York, 1979), 166.
92. S. Pickett and J. Thompson, "Patch Dynamics and the Design of Nature Reserves," *Biological Conservation* 13 (1978), 34.
93. H. Shugart and D. West, "Long-Term Dynamics of Forest Ecosystems," *American Scientist* 69 (1981), 647–52.
94. W. Romme and D. Despain, "Effects of Spatial Scale on Fire History and Landscape Dynamics in Yellowstone National Park." Paper presented at the Fifth Annual Landscape Ecology Symposium, March 21, 1990, Oxford, Ohio.
95. See in general F. Craighead, *Track of the Grizzly* (San Francisco, 1979).
96. Ibid., 157. In the North Cascades, Jon Almack uses 100 square miles as a rough average.
97. P. Bourgeron, "Advantages and Limitations of Ecological Classification for the Protection of Ecosystems," *Conservation Biology* 2 (1988), 219.
98. D. Sprugel, "Disturbance, Equilibrium, and Environmental Variability: What Is 'Natural' Vegetation in a Changing Environment?" *Biological Conservation* 6 (1991), 1–18; P. White, "Natural Disturbance, Patch Dynamics, and Landscape Pattern in Natural Areas," *Natural Areas Journal* 7 (1987), 14–22.
99. D. Sprugel, "Disturbance," 14.
100. R. Noss, "Corridors in Real Landscapes: A Reply to Simberloff and Cox," *Conservation Biology* 1 (1987), 159–64; D. Simberloff and J. Cox, "Consequences and Costs of Conservation Corridors," *Conservation Biology* 1 (1987), 63–71.
101. See A. Bennett, "Habitat Corridors and the Conservation of Small Mammals in a Fragmented Forest Environment," *Landscape Ecology* (in press).
102. As quoted in W. Stevens, "New Eye on Nature: The Real Constant is Eternal Turmoil," *New York Times,* July 31, 1990, B5.
103. R. Graham, "Response of Mammalian Communities to Environmental Changes during the Late Quaternary," in J. Diamond and T. Case, eds., *Community Ecology* (New York, 1986), 300–313.
104. M. Davis, "Climatic Instability, Time Lags, and Community Disequilibrium," in J. Diamond and T. Case, eds., *Community Ecology* (New York, 1986), 269.
105. D. Worster, "The Ecology of Chaos and Harmony," *Environmental History Review* 14 (1990), 13.
106. L. Harris and P. Gallagher, "New Initiatives for Wildlife Conservation:

The Need for Movement Corridors," in G. MacIntosh, ed., *Preserving Communities and Corridors* (Washington, D.C., 1989), 23.

107. R. Peters and J. Darling, "The Greenhouse Effect and Nature Reserves," *BioScience* 35 (1985), 707–17. See in general R. Peters and T. Lovejoy, eds., *Global Warming*.
108. As quoted in the *San Francisco Chronicle*, November 8, 1990, 25.
109. D. Botkin, *Discordant Harmonies* (New York, 1990), 190.
110. M. Hunter et al., "Paleoecology and the Coarse-Filter Approach to Maintaining Biological Diversity," *Conservation Biology* 2 (1988), 376.

Chapter 3: Ghost Bears

1. P. Shepard and B. Sanders, *The Sacred Paw* (New York, 1985), 2–6.
2. P. Sullivan, "A Preliminary Study of Historic and Recent Reports of Grizzly Bears, Ursus Arctos, in the North Cascades Area of Washington." Unpublished report of the Washington State Department of Game, 1983, p. 6.
3. IGBC, *Grizzly Bear Compendium* (Missoula, Montana, 1987), 3.
4. D. R. Wallace, *The Klamath Knot* (San Francisco, 1983), 3.
5. L. Brubaker, "Vegetation History and Anticipating Future Vegetative Change," in J. Agee and D. Johnson, eds., *Ecosystem Management for Parks and Wilderness* (Seattle, 1988), 50–51.
6. The Bear Mother myth is adapted from the version in G. Snyder, *The Practice of the Wild* (San Francisco, 1990), 155–61. See also Shepard and Sanders, *The Sacred Paw*, 55–59.
7. Shepard and Sanders, *The Sacred Paw*, 57.
8. J. Collins, *Valley of the Spirits* (Seattle, 1974), 17.
9. A. Hulkrantz, *The Religions of the American Indians* (Berkeley, 1979), 22; I. Paulsen, "The Preservation of Animal Bones in the Hunting Rites of Some Northern Eurasian Peoples," in V. Dioszegi, ed., *Popular Beliefs and Folklore Tradition in Siberia* (Bloomington, Indiana, 1968), 77.
10. Shepard and Sanders, *The Sacred Paw*, 108–10.
11. J. Craighead, J. Sumner, and G. Scaggs, *A Definitive System for Analysis of Grizzly Bear Habitat and Other Wilderness Resources* (Missoula, Montana, 1982).
12. IGBC, *Bear Tracks: Report* (Denver, 1989), 2.
13. J. Almack, *North Cascades Grizzly Bear Project: Annual Report* (Olympia, Washington, 1986), 10.
14. T. Barnard, "Current Population Status of Black Bear, Grizzly Bear, Cougar, and Wolf in the Skagit River Watershed," *British Columbia Ministry of the Environment Report* (Vancouver, British Columbia, 1986), 6.
15. F. Allendorf et al., "Estimation of Effective Population Size of Grizzly Bears by Computer Simulation," *Proceedings of the Fourth International Congress of Systematic and Evolutionary Biology*, (in press); see also F. Allendorf and C. Servheen, "Genetics and Conservation of Grizzly Bears," *Trends in Ecology and Evolution* 1: 4 (1986), 88–89.

16. M. Friedman, ed., *Forever Wild: Conserving the Greater North Cascades Ecosystem* (Bellingham, Washington, 1988), 62.
17. Ibid.
18. J. Almack, *Annual Report,* 10. See also B. Blanchard and R. Knight, "Movement of Yellowstone Grizzly Bears," *Biological Conservation* 58 (1991), 41–67.
19. B. McClellan and D. Shackleton, "Grizzly Bears and Resource Extraction Industries: Effects of Roads on Behavior, Habitat Use, and Demography," *Journal of Applied Ecology* 25 (1988), 457–60.
20. A. Dood, R. Brannon, and R. Mace, *Final Environmental Impact Statement: The Grizzly Bear in Northwest Montana* (Helena, Montana, 1986), 22.
21. D. Mattson, R. Knight, and B. Blanchard, "The Effects of Development and Primary Roads on Grizzly Bear Habitat Use in Yellowstone National Park, Wyoming," *International Conference on Bear Research and Management* 7 (1987), 259–73.
22. W. Archibald, R. Ellis, and A. Hamilton, "Responses of Grizzly Bears to Logging Truck Traffic in the Kionsquit River Valley, B.C.," *International Conference on Bear Research and Management* 7 (1987), 251–57.
23. W. Kasworm and T. Manley, "Road and Trail Influences on Grizzly Bears and Black Bears in Northwest Montana," *International Conference on Bear Research and Management* 8 (in press).
24. Ibid.
25. As quoted in J. Mills, "The Grizzly and the Hand of Man," *American Forests* (January–February 1989), 28.
26. U.S. Forest Service, *Land and Resource Management Plan: Wenatchee National Forest* (Portland, 1986), 4: 82.
27. U.S. Forest Service, *Land and Resource Management Plan: Okanogan National Forest* (Portland, 1986), 3: 69.
28. U.S. Forest Service, *Land and Resource Management Plan: Mt. Baker–Snoqualmie Forest* (Portland, 1990), 4: 68–9.
29. P. Lee and J. Weaver, "Biological Evaluation of Man/Grizzly Bear Conflicts Related to Sheep Grazing in Essential Grizzly Bear Habitat on the Targhee National Forest" (St. Anthony, Idaho, 1981), 7.
30. U.S. Forest Service, *Early Winters Alpine Winter Sports Final EIS* (Portland, 1984), 37.
31. S. Herrero, *Bear Attacks: Their Causes and Avoidance* (Piscataway, New Jersey, 1985), 187.
32. B. J. Williams, "Spectre of the North Cascades Grizzly," *Washington Magazine* (September–October 1989), 47–50.
33. Almack, *Annual Report,* 13.
34. Ibid., 35.
35. U.S. Fish and Wildlife Service, *Draft Revised Grizzly Bear Recovery Plan* (Missoula, Montana, 1990).

36. As quoted in K. Hammer, "Charting the Course to Extinction," *Earth First! Journal* (December 21, 1990), 26.
37. As quoted in K. Hammer, "Grizzlies at Risk," *Forest Watch* 11: 6 (1990), 23.
38. K. Hammer, "Charting the Course," 27.
39. Ibid.
40. U.S. Fish and Wildlife Service, *Draft Revised Grizzly Bear Recovery Plan*, 12.
41. A Christensen and M. Madel, "Cumulative Effects Analysis Process: Grizzly Bear Habitat Component Mapping" (USDA Forest Service, Libby, Montana, 1982), 26.
42. C. Servheen, personal communication.
43. J. Almack, "Draft Grizzly Bear Workbook," April 1988. Unpublished report, available from the author.
44. Data available from the author.
45. J. Almack, North Cascades Bear Investigations: 1989 Progress Report (Olympia, Washington, 1989), 18.
46. From J. Haas, U.S. Fish and Wildlife Service, Olympia, Washington, personal communication.
47. Shepard and Sanders, *The Sacred Paw*, 204.

Chapter 4: Laws on the Land

1. J. Laufer and P. Jenkins, *A Preliminary Study of Gray Wolf History and Status in the Region of the Cascade Mountains of Washington State* (Tenino, Washington, 1989), 5.
2. R. Culbert and R. Blair, "Recovery Planning and Endangered Species," *Endangered Species Update* 6: 10 (1989), 2.
3. USDI Office of the Inspector General, *Audit Report: The Endangered Species Program, U.S. Fish and Wildlife Service* (Washington, D.C., 1990), appendix 1.
4. K. Franklin, "Endangered Species: Where to Go from Here?" *American Forests* (November/December 1987), 57.
5. *Statutes at Large of the United States of America, 1789–1873* (Washington, D.C., 1850–73), 17 (1872): 32–33.
6. A. Runte, *National Parks: The American Experience* (Lincoln, Nebraska, 1987), 194.
7. *United States Statutes at Large* (Washington, D.C., 1874–), 39 (1916): 535.
8. Ibid., 48 (1934): 817.
9. G. Coggins, "Protecting the Wildlife Resources of National Parks from External Threats," *Land and Water Review* 22 (1987), 16–18; J. Sax and R. Keiter, "Glacier National Park and Its Neighbors: A Study of Federal Interagency Relations," *Ecology Law Quarterly* 14 (1987), 207–63; W. Lockhart, "External Park Threats and Interior's Limits: The Need for an

Independent Park Service," in D. Simon, ed., *Our Common Lands: Defending the National Parks* (Washington, D.C., 1988), 3–72.

10. "Report Language: Task Directive on Park Study," undated memo from D. Fluharty, director, North Cascades Conservation Council, 4–5.
11. U.S. Code 16, sec. 1131–36 (1982).
12. R. Noss, "What Can Wilderness Do for Biodiversity?" *Wild Earth* 1 (1991), 54.
13. D. Crumpacker, S. W. Hodge, D. Friedley, and W. P. Gregg, Jr., "A Preliminary Assessment of the Status of Major Terrestrial and Wetland Ecosystems on Federal and Indian Lands in the U.S.," *Conservation Biology* 2 (1988), 103–15.
14. See, in general, M. Bean, *The Evolution of National Wildlife Law* (Washington, D.C., 1977), 66–261, 288–319.
15. S. Yaffee, *Prohibitive Policy: Implementing the Federal Endangered Species Act* (Cambridge, Massachusetts, 1982), 188.
16. Ibid., 13.
17. For a related discussion, see D. Rohlf, "Six Biological Reasons Why the Endangered Species Act Doesn't Work—and What to Do about It," *Conservation Biology* 5: 3 (1991), 273–82; M. O'Connell, "Response to: 'Six Reasons Why the Endangered Species Act Doesn't Work and What to Do About It,' "*Conservation Biology* 6 (1992) 140–43; D. Rohlf, "Response to O'Connell," *Conservation Biology* 6 (1992), 144–45. See also H. Doremus, "Patching the Ark: Improving Legal Protection of Biological Diversity," *Ecology Law Quarterly* 18: 21 (1991), 265–333.
18. U.S. Code sec. 16, 1533 (f)(1)(A) (1988).
19. U.S. Fish and Wildlife Service, "Federal and State Endangered Species Expenditures in Fiscal Year 1990," *Endangered Species Technical Bulletin* 16: 5 (1991), 3.
20. See in general, M. Soulé, ed., *Viable Populations for Conservation* (Cambridge, Massachusetts, 1987). Subpopulations can be difficult to identify as taxonomic studies become more refined. For a discussion, see H. Shaw, *Soul among Lions* (Boulder, 1990), 104–10.
21. J. Sidel and D. Bowman, "Habitat Protection under the Endangered Species Act," *Conservation Biology* 2 (1988), 116.
22. U.S. Code 16, sec. 1532 (5)(A) (1988).
23. D. Rohlf, *The Endangered Species Act* (Stanford, California, 1989), 57. See also P. Salzman, "Evolution and Application of Critical Habitat under the Endangered Species Act," *Harvard Environmental Law Review* 14 (1990), 311, 322–23; Yagerman, "Protecting Critical Habitat under the Federal Endangered Species Act," *Environmental Law* 20 (1990), 811–55.
24. Ibid, 51.
25. Ibid., 107.
26. Ibid., 132.
27. G. Meese, "Saving Endangered Species," in G. MacIntosh, ed., *Preserving Communities and Corridors* (Washington, D.C., 1989), 53.

28. See in general U.S. General Accounting Office, *Endangered Species: Management Improvements Could Enhance Recovery Program* (Washington, D.C., 1988).
29. R. Culbert and R. Blair, "Recovery Planning," 3–4, 6.
30. J. Fitzgerald, "The 1988 Recovery Amendment: Its Evolution and Content," *Endangered Species Update* 7: 1 and 2 (1989), 2. Of course, there are political compromises made *throughout* the process—see in general U.S. General Accounting Office, *Endangered Species: A Controversial Issue Needing Resolution* (Washington, D.C., 1979).
31. B. Parry, "Cumulative Habitat Loss: Cracks in the Environmental Review Process," *Natural Areas Journal* 10: 2 (1990), 76.
32. U.S. Code 16, sec. 1604 (g)(3)(B) (1982).
33. See in general, C. Wilkinson and H. Anderson, *Land and Resource Planning in the National Forests* (Washington, D.C., 1987).
34. Ibid., 71.
35. Ibid., 280–90.
36. Ibid., 292.
37. U.S. Code 16, sec. 1604 (g)(3)(B) (1982).
38. Wilkinson and Anderson, *Land and Resource Planning,* 170–73.
39. *Final Report of Committee of Scientists,* 44 Fed. Reg. at 26, 609 (1979).
40. Code of Federal Regulations, title 36, sec. 219.27 (g) (1984); sec. 219.1 (b)(3) (1984); and sec. 219.26 (1984).
41. Ibid., sec. 219.3 (1984).
42. Ibid., sec. 219.19 (1984).
43. For a critique of this theory, see P. Landres, J. Verner, and J. W. Thomas, "Ecological Uses of Vertebrate Indicator Species: A Critique," *Conservation Biology* 2 (1988), 316–28.
44. Code of Federal Regulations, title 36, sec. 219.19 (a)(2) (1984).
45. Wilkinson and Anderson, *Land and Resource Planning,* 300.
46. Code of Federal Regulations, 36 sec. 219.26 (1984).
47. Ibid., sec. 219.12 (d) (1984).
48. Ibid., sec. 219.19 (a)(6) (1984).
49. *Final Report of the Committee of Scientists,* at 26, 608.
50. The Wilderness Society, *Forests for the Future? A Report on National Forest Planning* (Washington, D.C., 1987).
51. The Wilderness Society, *End of the Ancient Forests* (Washington, D.C., 1988), 31–35.
52. J. F. Torrence, *Diversity in Forest Plans* (Portland, 1988), memo 2620/1920, p. 1.
53. U.S. Forest Service, *Record of Decision, Land and Resource Management Plan, Mt. Baker–Snoqualmie National Forest* (Portland, 1990), 12–14.
54. The following are examples: L. Harris, *The Fragmented Forest* (Chicago, 1984); K. Rosenberg and M. Raphael, "Effects of Forest Fragmentation on Vertebrates in Douglas-Fir Forests," in J. Verner, M. Morrison, and J. Ralphs, eds., *Wildlife 2000: Modelling Habitat Relationships of Terres-*

trial Vertebrates (Madison, Wisconsin, 1986), 263–72; B. Wilcox and D. Murphy, "Conservation Strategy: The Effects of Fragmentation on Extinction," *American Naturalist* 125 (1985), 879–87; D. Wilcove et al., "Habitat Fragmentation in the Temperate Zone," in M. Soulé, ed., *Conservation Biology: The Science of Scarcity and Diversity* (Sunderland, Massachusetts, 1986), 237–56.

55. P. Morrison, *Old Growth in the Pacific Northwest* (Washington, D.C., 1988), 39.
56. See in general, A. Bennett, "Roads, Roadsides, and Wildlife Conservation," in D. Saunders and R. Hobbs, eds., *Nature Conservation: The Role of Corridors* (Chipping Norton, NSW, Australia, 1991); and Earth First! Biodiversity Task Force, *Killing Roads: A Citizen's Primer on the Effects and Removal of Roads* (Tucson, 1990).
57. J. Franklin and R. T. T. Forman, "Creating Landscape Patterns by Forest Cutting: Ecological Consequences and Principles," *Landscape Ecology* 1 (1987), 5–18; see in general L. Harris, *The Fragmented Forest.*
58. U.S. Forest Service, *Land and Resource Management Plan, Mt. Baker–Snoqualmie National Forest,* 3: 93.
59. Greater Ecosystem Alliance, *Appeal of the Land and Resource Management Plan, Mt. Baker–Snoqualmie National Forest* (Bellingham, Washington, 1990), 17.
60. Sierra Club, John Muir Chapter, and Wisconsin Forest Conservation Task Force, "Appeal of the Land and Resource Management Plan, Chequamegon National Forest," before the chief, U.S. Forest Service, August 11, 1986.
61. M. Soulé, "Thresholds for Survival: Maintaining Fitness and Evolutionary Potential," in M. Soulé and B. Wilcox, eds., *Conservation Biology: An Evolutionary-Ecological Perspective* (Sunderland, Massachusetts, 1980), 151–69.
62. Sierra Club, John Muir Chapter, "Appeal," 83.
63. Ibid., 2.
64. W. Kuhlmann, "A Biological Attack on Timber Primacy," *Forest Watch* 11: 1 (1990), 15. See also W. Kuhlmann, "The Wisconsin Biodiversity Lawsuit," *Inner Voice* 3: 6 (1991), 10.
65. Code of Federal Regulations, title 36, sec. 219.26 (1984).
66. W. Kuhlmann, "A Biological Attack," 21.
67. R. Keiter, "Taking Account of the Ecosystem on the Public Domain: Law and Ecology in the Greater Yellowstone Region," *University of Colorado Law Review* 60 (1989), 965.
68. Ibid., 964.
69. R. O'Toole, "Final 1990 RPA Program Reduces Timber Sales," *Forest Watch* 11: 1 (1990), 14.
70. R. Simmons, "An Exercise in Futility," *Forest Watch* 10: 1 (1989), 30.
71. R. O'Toole, "Back to the Drawing Board," *Forest Watch* 10: 1 (1989), 24.
72. G. Studds, opening statement, Subcommittee on Fisheries and Wildlife

Conservation and the Environment Hearing on H.R. 1268, 100th Cong., 2d sess., November 16, 1989, p. 1.

73. H. Salwasser, "Concerning the Forest Service's Role in Conserving Biological Diversity," statement before the Subcommittee on Fisheries and Wildlife Conservation and the Environment Hearing on H.R. 1268, 100th Cong., 2d sess., November 16, 1989, pp. 6, 8.
74. L. Blumenthal, "Forest Fallout: Industry Changes, Not Owl, Blamed," *Tacoma News Tribune,* August 18, 1991.
75. K. N. Johnson et al., "Alternatives for Management of Late-Successional Forests of the Pacific Northwest," A report to the Agriculture Committee and the Merchant Marine and Fisheries Committee of the U.S. House of Representatives, 102d Cong., 2d sess., October 8, 1991, p. 12.
76. J. Jontz, Ancient Forest Protection Act of 1991 (H.R. 842), introduced to the U.S. House of Representatives, 102d Cong., 1st sess., February 6, 1991, p. 4.
77. B. Vento, Ancient Forests Act of 1990 (H.R. 5295), introduced into the U.S. House of Representatives, 101st Cong., 2d sess., July 18, 1990, pp. 5–6. The 1991 version has the same goals.
78. B. Adams, Pacific Northwest Forest Community Recovery and Ecosystem Conservation Act of 1991 (S. 1536), introduced into the U.S. Senate, 102d Cong., 1st sess., July 24, 1991, p. 5.
79. As quoted in "The Battle Brews," *Forest Watch* 11: 10 (1991), 13.
80. Report of the Task Force on Endangered Species, National Conference on Forest Service Reform, Camp Cascade, Oregon, September 14, 1990, p. 71.
81. R. Keiter, "Taking Account of the Ecosystem," 997.
82. G. Meese, "Saving Endangered Species," 55.
83. E. Diringer, "Protection for Forests Broadened," *San Francisco Chronicle,* September 14, 1990, p. 1.

Chapter 5: The Landscape of Management

1. UNESCO, *Use and Preservation of the Biosphere: Proceedings of the Intergovernmental Conference of Experts on the Scientific Basis for Rational Use and Conservation of the Resources of the Biosphere* (Paris, 1968), 16.
2. C. Maser, *The Redesigned Forest* (San Pedro, California, 1988), 59.
3. For thorough reviews see in general D. Ehrenfeld, *The Arrogance of Humanism* (New York, 1978); N. Evernden, *The Natural Alien* (Toronto, 1985); M. Berman, *The Reenchantment of the World* (Ithaca, New York, 1981); C. Merchant, *The Death of Nature: Women, Ecology, and the Scientific Revolution* (San Francisco, 1980); R. Nash, *The Rights of Nature* (Madison, Wisconsin, 1989); B. Devall and G. Sessions, *Deep Ecology* (Salt Lake City, 1985); and M. Oelschlager, *The Idea of Wilderness* (New Haven, 1991).
4. C. Allin, *The Politics of Wilderness Preservation* (Westport, Connecticut, 1982), 7–8.

5. R. Nash, *Wilderness and the American Mind* (New Haven, 1983), 24–25.
6. C. Allin, *Wilderness Preservation,* 12.
7. Ibid., 19.
8. R. Nash, *Wilderness,* 96–107.
9. C. Allin, *Wilderness Preservation,* 34; J. Ise, *United States Forest Policy* (New Haven, 1920), 118.
10. Valuable general treatments of evolving U.S. land-management attitudes are C. Allin, *Wilderness Preservation;* R. Nash, *Wilderness;* J. Ise, *Our National Park Policy: A Critical History* (Baltimore, 1961); S. Dana and S. Fairfax, *Forest and Range Policy* (New York, 1979); and H. Steen, *The U.S. Forest Service: A History* (Seattle, 1976).
11. R. Nash, *Wilderness,* 129.
12. G. Pinchot, *Breaking New Ground* (New York, 1947), 261.
13. Ibid., 326.
14. Ibid., 261.
15. D. Snow, "Save the Forests, Store the Floods," in *Northern Lights* 7: 3 (1991), 12–17.
16. D. Botkin, *Discordant Harmonies* (New York, 1990), 56–68.
17. Editorial, *Journal of Forestry* 15 (1943), 25–26.
18. C. Koppes, "Efficiency/Equity/Esthetics: Towards a Reinterpretation of American Conservation," *Environmental Review* 11: 2 (1987), 127.
19. See S. Hays, *Conservation and the Gospel of Efficiency: The Progressive Conservation Movement, 1890–1920* (Cambridge, Massachusetts, 1959).
20. For an excellent biography of Muir, see S. Fox, *The American Conservation Movement: John Muir and His Legacy* (Madison, Wisconsin, 1981). See also M. Cohen, *The Pathless Way* (Madison, 1984).
21. W. Fox, "The Deep Ecology–Ecofeminism Debate and Its Parallels," *Environmental Ethics* 2 (1989), 23.
22. D. Zaslowsky, *These American Lands* (New York, 1986), 20, 78.
23. S. Dana and S. Fairfax, *Forest and Range Policy,* 209.
24. H. Steen, *The U.S. Forest Service,* 118–22; S. Dana and S. Fairfax, *Forest and Range Policy,* 155; C. Allin, *Wilderness Preservation,* 71; C. Allin, "Wilderness Preservation as a Bureaucratic Tool," in P. Foss, ed., *Federal Lands Policy* (New York, 1987), 131–32.
25. K. Van Tighem, "Have Our National Parks Failed Us?" *Park News* (1986), 31–33.
26. W. Berry, "Preserving Wildness," in *Home Economics* (San Francisco, 1987), 139.
27. D. Hales, "Changing Concepts of Parks," in D. Western and M. Pearl, eds., *Conservation for the Twenty-First Century* (New York, 1989) 143.
28. P. Ehrlich, D. Dobkin, and D. Wheye, *The Birders Handbook* (New York, 1988), 291, 293.
29. For an overview see M. Bonnett and K. Zimmerman, "Politics and Preser-

vation: The Endangered Species Act and the Northern Spotted Owl," *Ecology Law Quarterly* 18: 1 (1991), 105–71.

30. D. Wilcove, "Of Owls and Ancient Forests," in E. Norse, *Ancient Forests of the Pacific Northwest* (Washington, D.C., 1990), 78. See also E. Forsman et al. (The Wildlife Society), *Distribution and Biology of the Spotted Owl in Oregon,* Wildlife Monographs no. 87 (1984), 16; U.S. Fish and Wildlife Service, *The Northern Spotted Owl Status Review Supplement* (Portland, 1989), 60, 2.10–2.11; M. Bonnett and K. Zimmerman, "Politics and Preservation," 111–14.
31. W. Dawson et al., "Report of the Scientific Advisory Panel on the Spotted Owl," *Condor* 89 (1987), 209.
32. D. Wilcove, "Owls and Ancient Forests," 80.
33. B. Marcot and R. Holthausen, "Analyzing Population Viability of the Spotted Owl in the Pacific Northwest," *Transactions of the North Wildlife Natural Resources Conference* 52 (1987), 333–47.
34. See in general U.S. Forest Service, *Regional Guide for Pacific Northwest Region* (Portland, 1984).
35. W. Dawson et al., "Report of the Scientific Advisory Panel," 220.
36. See in general U.S. Forest Service, *Final Supplement to the Environmental Impact Statement for an Amendment to the Pacific Northwest Regional Guide*, vols. 1–2 (Portland, 1988).
37. Ibid., vol. 4, 34.
38. U.S. Government Accounting Office, *Endangered Species Spotted Owl Petition Beset by Problems* (Washington, D.C., 1989), 1.
39. U.S. Fish and Wildlife Service, "Listing Action Completed for Spotted Owl and Five Other Species," *Endangered Species Technical Bulletin* 15: 7 (1990), 5.
40. J. Thomas et al., *Conservation Strategy,* 4.
41. National Audubon Society, *Notes on the Thomas Committee Report and Protection of the Spotted Owl* (Olympia, Washington, 1990), 1–4.
42. J. Thomas et al., *Conservation Strategy,* 4.
43. D. Doak, "Spotted Owls and Old Growth Logging in the Pacific Northwest," *Conservation Biology* 3 (1989), 395. See also R. Lande, "Demographic Models of the Northern Spotted Owl," *Oecologia* 75 (1988).
44. As quoted in "Owl Calls," *Forest Watch* 11: 3 (1990), 8.
45. Quoted in J. Stiak, "Bush Team Quick-Kicks the Spotted Owl to Congress," *High Country News* (July 16, 1990), 5.
46. Quoted in "Owl Calls," 6.
47. Quoted in "Lujan to Cut Back Job Losses in Owl Plan," *Seattle Post-Intelligencer,* August 27, 1990, p. 1.
48. Quoted in J. Stiak, "Bush Team Quick-Kicks the Spotted Owl," 5.
49. Quoted in Cascade Holistic Economic Consultants staff, "More Owls and Jobs?" *Forest Watch* 10: 10 (1990), 16.

50. "Ninth Circuit Court Strikes Down Section 318's Restrictions on Judicial Review as Unconstitutional," *Forest Watch* 11: 3 (1990), 25.
51. As quoted in "Judge Dwyer Does It Again," *Forest Watch* 11: 10 (1991), 10–11.
52. Statement of Kika de la Garza, chairman, House Agriculture Committee on the Final Scientific Report on Options to Protect Old-Growth Forests, 102d Cong., 2d. sess., Tuesday, October 8, 1991.
53. As quoted in "New Alternatives for the Pacific Northwest?" *Forest Watch* 11: 11 (1991), 9.
54. K. N. Johnson et al. "Alternatives for Management of Late-Successional Forests of the Pacific Northwest." A report to the Agriculture Committee and the Merchant Marine and Fisheries Committee of the U.S. House of Representatives, 102d Cong., 2d sess., October 8, 1991, p. 8.
55. As quoted in T. Abate, "Four Scientists Analyze Ancient Forests for a Congressional Committee," *High Country News* (October 21, 1991), 8.
56. K. N. Johnson et al., "Alternatives for Management," 15.
57. H. Salwasser, "Spotted Owls: Turning a Battleground into Blueprint," *Ecology* 68 (1987), 777.
58. J. Franklin, "The Biosphere Reserve Program in the United States," *Science* 195 (1977), 262–67; M. Dyer and M. Holland, "UNESCO's Man and the Biosphere Program," *BioScience* 38 (1988), 635–42; J. Hough, "Biosphere Reserves: Myth and Reality," *Endangered Species Update* 6 (1988), 1–4; L. Tangley, "A New Era for Biosphere Reserves," *BioScience* 38 (1988), 148–55; M. Dyer and M. Holland, "The Biosphere Reserve Concept: Needs for a Network Design," *BioScience* 41 (1991), 319–25.
59. J. Hough, "Biosphere Reserves," 2.
60. M. Dyer and M. Holland, "UNESCO's Biosphere Program," 637.
61. See in general J. Agee and D. Johnson, *Ecosystem Management for Parks and Wilderness* (Seattle, 1988).
62. T. Clark and A. Harvey, *Management of the Greater Yellowstone Ecosystem: An Annotated Bibliography* (Jackson, Wyoming, 1988), 7–24.
63. T. Clark and D. Zaunbrecher, "The Greater Yellowstone Ecosystem: The Ecosystem Concept in Natural Resource Policy and Management," *Renewable Resources Journal* (summer 1987), 11.
64. R. Baker and C. Schonewald-Cox, "Management Strategies for Improving Population Viability," in B. Wilcox, P. Brussard, and B. Marcot, eds., *The Management of Viable Populations: Theory, Applications, and Case Studies* (Stanford, 1987), 73–87.
65. H. Salwasser, C. Schonewald-Cox, and R. Baker, "The Role of Interagency Cooperation in Managing for Viable Populations," in M. Soulé, ed., *Viable Populations for Conservation* (New York, 1987), 147–73.
66. R. Keiter, "NEPA and the Emerging Concept of Ecosystem Management on the Public Lands," *Land and Water Law Review* 25: 1 (1990), 48.

67. "Why Public Opinion Never Reaches the Top," *The Inner Voice* 2: 4 (1990), 15.
68. As quoted in T. Ribe, "To ASQ or Not to ASQ," *The Inner Voice* 2: 4 (1990), 15.
69. See in general B. Twight, *Organizational Values and Political Power: The Forest Service vs. Olympic National Park* (University Park, Pennsylvania, 1983); P. Mohai, "Public Participation and Natural Resource Decision-Making," *Natural Resources Journal* 27 (1987), 123–55.
70. See in general R. O'Toole, *Reforming the Forest Service* (Washington, D.C., 1988).
71. P. Culhane, *Public Lands Politics* (Baltimore, Maryland, 1981), 334–36; see in general G. McConnell, *Private Power and American Democracy* (New York, 1967) and D. Clary, *Timber and the Forest Service* (Lawrence, Kansas, 1986).
72. See A. Miller, "Technological Thinking: Its Impact on Environmental Management," *Environmental Management* 9: 3 (1985), 179–90; T. Clark, "Professional Excellence in Wildlife and Natural Resources Organizations," *Renewable Resources Journal* 4: 3 (1986), 8–13; S. Yaffee, *Prohibitive Policy: Implementing the Endangered Species Act* (Cambridge, Massachusetts, 1982), 9–10; J. Kennedy, "Legislative Confrontation of Groupthink in U.S. Natural Resource Agencies," *Environmental Conservation* 15: 2 (1988), 123–28; P. Mohai, "Public Participation," 123–55; R. Andrews, "Values Analysis in Environmental Policy," in D. Mann, ed., *Environmental Policy Formation* (Lexington, Massachusetts, 1986), 137–47; P. Culhane, *Public Lands Politics,* 325.
73. J. Wiebush, "Bureaucratic Gridlock Demoralizing USFS," *The Inner Voice* 2: 4 (1990), 7.
74. Congressional Research Service, *Greater Yellowstone Ecosystem: An Analysis of Data Submitted by Federal and State Agencies* (Washington, D.C., 1987), 9–11. Similar conclusions were reached more recently. See D. Glick et al., eds., *An Environmental Profile of the Greater Yellowstone Ecosystem* (Bozeman, Montana, 1991), 98–108; see also in general R. Keiter and M. Boyce, *The Greater Yellowstone Ecosystem* (New Haven, 1991).
75. Congressional Research Service, *Greater Yellowstone Ecosystem*, 9–11.
76. R. Keiter, "Taking Account of the Ecosystem on the Public Domain: Law and Ecology in the Greater Yellowstone Region," *University of Colorado Law Review* 60 (1989), 987, note 313.
77. Lorraine Mintzmeyer, testimony before the U.S. House of Representatives Civil Service Subcommittee, September 24, 1991, as quoted in *High Country News* 23: 18 (1991), 10.
78. Greater Yellowstone Coordinating Committee, "Draft Vision for the Future: A Framework for Coordination in the Greater Yellowstone Area." (Billings, Montana, August 1990), 1–1, 3–1.

79. As quoted in T. Williams, "All Opposed to the Future," *Fly Rod and Reel Magazine* (July–October 1991), 17.
80. Quoted in ibid.
81. Ibid.
82. Greater Yellowstone Coordinating Committee, "A Framework for Coordination of National Parks and National Forests in the Greater Yellowstone Area," (Billings, September 1991), 1.
83. Mintzmeyer, Testimony, 10–11.
84. U.S. Code 16, sec. 1604 (a) (1982).
85. U.S. Forest Service, *Land and Resource Management Plan, Mt. Baker–Snoqualmie National Forest* (Portland, 1990), A-10.
86. R. Keiter, "Taking Account of the Ecosystem," 989.
87. U.S. Park Service, *General Management Plan, North Cascades National Park* (Seattle, 1988), 2.
88. John Reynolds, former superintendent, North Cascades National Park, personal communication.
89. J. Sax and R. Keiter, "Glacier National Park and Its Neighbors: A Study of Federal Interagency Relationships," *Ecology Law Quarterly* 14 (1987), 207–63; R. Keiter, "Natural Ecosystem Management in Parks and Wilderness Areas: Looking at the Law," in J. Agee and D. Johnson, eds., *Ecosystem Management for Parks and Wilderness,* 15–40; V. Gilbert, "Cooperation in Ecosystem Management," in J. Agee and D. Johnson, eds., *Ecosystem Management for Parks and Wilderness,* 180–92.
90. See in general U.S. Forest Service, *Land and Resource Management Plan, Cibola National Forest* (Albuquerque, 1985).
91. K. Matthews, "Hard Negotiating or Treachery?" *High Country News* (July 2, 1990), 13.
92. R. O'Toole, "The Case for Repealing the NFMA," *Forest Watch* (January–February 1990), S-2.
93. "Why Public Opinion Never Reaches the Top," 6.
94. V. Gilbert, "Cooperation in Ecosystem Management," 182.
95. G. Leonard, "Joining Forces to Manage Our Old Growth Forests," *Forest Watch* (December 1989), 16.
96. AFSEEE staff, *Inner Voice* 1 (1989), 1.
97. P. Hirt, "Dissension within the Ranks," *Earth First! Journal* (May 1, 1990), 7.
98. Ibid.
99. John Mumma, testimony before the U.S. House of Representatives Civil Service Subcommittee, September 24, 1991. As quoted in *High Country News* 23: 18 (1991), 12.
100. Quoted in J. St. Clair, "Was Overbay's Powerplay an Overreach?" *Forest Watch* 12: 4 (1991), 14.
101. J. St. Clair, "The Lolo Goes Solo," *Forest Watch* 2: 3 (1991), 14.
102. Quoted in J. St. Clair, "Owl Calls: Cutting by Numbers," *Forest Watch* 12: 4 (1991), 9.

103. Ibid., 10.
104. Larry Harris as quoted in D. Chadwick, "The Biodiversity Challenge," 11.
105. See J. B. Callicott, "Standards of Conservation: Then and Now," *Conservation Biology* 4 (1990), 230–31.
106. W. Berry, *What Are People For?* (San Francisco, 1990), 13.

Chapter 6: Ecosystem Management for Native Diversity

1. B. Norton, "Operationalizing the Land Ethic: Toward an Integrated Theory of Environmental Management," undated manuscript, 19; available from author at Social Sciences Division, Georgia Institute of Technology, Atlanta, Georgia, 30332. See E. Bird, "The Social Construction of Nature: Theoretical Approaches to the History of Environmental Problems," *Environmental Review* 11 (1987), 255–64.
2. R. Rappaport, "The Flow of Energy in an Agricultural Society," *Scientific American* 224 (1971), 132.
3. J. Karr and D. Dudley, "Ecological Perspective on Water Quality Goals," *Environmental Management* 5 (1981), 56.
4. J. Karr et al., "Assessing Biological Integrity in Running Waters: A Method and Its Rationale," *Special Publication 5, Illinois Natural History Survey* (Urbana, Illinois, 1986).
5. B. Norton, "Operationalizing the Land Ethic," 11.
6. E. Grumbine, "Protecting Biological Diversity through the Greater Ecosystem Concept," *Natural Areas Journal* 10: 3 (1990), 114–20.
7. B. Norton, "What Is a Conservation Biologist?" *Conservation Biology* 2 (1988), 238. For a full treatment of environmental economics, see H. Daley and J. Cobb, *For the Common Good* (Boston, 1990).
8. R. Noss, "Issues of Scale in Conservation Biology," in P. Fiedler and S. Jain, eds., *The Theory and Practice of Nature Conservation, Preservation, and Management* (New York, in press).
9. M. Scott et al., "Beyond Endangered Species: An Integrated Conservation Strategy for the Preservation of Biological Diversity," *Endangered Species Update* 5: 10 (1988), 43–48.
10. See in general M. Soulé and K. Kohm, eds., *Research Priorities for Conservation Biology* (Washington, D.C., 1989).
11. H. Howe and S. Hubbell, "Progress Report on Proposing a National Institute For the Environment," *BioScience* 40 (1990), 567.
12. R. Noss and L. Harris, "Nodes, Networks, and MUM's: Preserving Diversity at All Scales," *Environmental Management* 10 (1986), 299–309.
13. R. Noss, "Wilderness Recovery: Thinking Big in Restoration Ecology," *The Environmental Professional* 13 (1991), 227.
14. R. Peters and J. Darling, "The Greenhouse Effect and Nature Reserves," *BioScience* 35 (1985), 707–17. See also in general R. Peters and T. Lovejoy, eds., *Global Warming and Biological Diversity* (New Haven, 1992).

15. See in general D. Smith, *The Practice of Silviculture* (New York, 1986); C. Maser, *The Redesigned Forest* (San Pedro, California, 1988).
16. J. Franklin, "Old Growth: The Contribution to Commercial Forests," *Forest Planning Canada* 5: 3 (1989), 17.
17. Ibid.; see in general J. Franklin et al., *Maintaining Long-Term Productivity of Pacific Northwest Forest Ecosystems* (Portland, 1989).
18. G. Davis, "Design of a Long-Term Ecological Monitoring Program for Channel Islands National Park, California," *Natural Areas Journal* 9: 2 (1989), 80–90.
19. P. and A. Ehrlich, *Extinction* (New York, 1981), 86–100.
20. R. Noss, "On Characterizing Presettlement Vegetation: How and Why," *Natural Areas Journal* 5: 1 (1985), 5–13.
21. W. Jordan et al., "Ecological Restoration as a Strategy for Conserving Biological Diversity," *Environmental Management* 12: 1 (1988), 55.
22. See in general U.S. Park Service, *Tenth Annual Report to Congress on the Status of Implementation of the Redwood National Park Expansion Act of 1978* (Crescent City, California, 1987).
23. G. Miller, *Living in the Environment* (Belmont, California, 1990), 378.
24. R. Keiter, "Taking Account of the Ecosystem on the Public Domain: Law and Ecology in the Greater Yellowstone Region," *University of Colorado Law Review* 60 (1989), 932.
25. E. Marston, "Metamorphosis at the Forest Service," *High Country News* (October 8, 1990), 14.
26. T. Clark, "Conservation Biology of the Black-footed Ferret," *Wildlands Preservation Trust International Species Science Report 3* (1989), 3.
27. R. Reading et al., "Towards an Endangered Species Reintroduction Paradigm," *Endangered Species Update* 8: 11 (1991), 1–4; T. Clark, "Practicing Natural Resource Management with a Policy Orientation," *Environmental Management* 16: 2 (1992).
28. P. Vitousek et al., "Human Appropriation of the Products of Photosynthesis," *BioScience* 36 (1986), 368–73.
29. D. Waller, "Sharing Responsibility for Conserving Diversity: The Complementary Roles of Conservation Biologists and Public Lands Agencies," *Conservation Biology* 2 (1988), 398–401; J. Thomas and H. Salwasser, "Bringing Conservation Biology into a Position of Influence in Natural Resource Management," *Conservation Biology* 3 (1989), 123–27.
30. J. Franklin et al., "Contributions of the Long-Term Ecological Research Program," *BioScience* 40 (1990), 509–23.
31. B. Jennings, "Representation and Participation in the Democratic Governance of Science and Technology," in M. Goggin, ed., *Governing Science and Technology in a Democracy* (Knoxville, Tennessee, 1986), 240.
32. Memo from William Mott, director, National Park Service, September 13, 1988, p. 1.
33. A. Boulton, ed., *Forest Watch,* special edition on forest planning, 10: 1 (1989); P. Culhane, *Public Lands Politics* (Baltimore, 1981), 325–27.

34. S. McCool et al., "An Alternative to Rational-Comprehensive Planning: Transactive Planning," in R. Lucas, comp., *Proceedings: National Wilderness Research Conference* (Ogden, Utah, 1986), 544. See also D. Henning, "Wilderness Politics: Public Participation and Values," *Environmental Management* 11 (1987), 285.
35. J. Ashor et al., "Improving Wilderness Planning Efforts: Application of the Transactive Planning Approach," in R. Lucas, *National Wilderness Research Conference,* 424–32; P. Bobrow et al., "Regional Planning Acceptance by Residents of Northern New York, U.S.A.," *Environmental Management* 8: 1 (1984), 45–54; J. Glass, "Citizen Participation in Planning: The Relationship between Objectives and Techniques," *American Planning Association Journal* 45: 3 (1979), 180–89; S. Grabow and A. Heskin, "Foundations for a Radical Concept of Planning," *Journal of the American Institute of Planning* 39: 2 (1973), 106.
36. E. Grumbine, "Can Wilderness Be Saved from Vibram Soles?" *High Country News* (May 27, 1985), 14–15.
37. G. Stokes, "LAC Task Force Role," in R. Lucas, *National Wilderness Research Conference,* 546. See also J. Friedmann, "The Public Interest and Community Participation: Toward a Reconstruction of Public Philosophy," *Journal of the American Institute of Planners* 39: 1 (1973), 2–12; S. McCool et al., "An Alternative to Rational-Comprehensive Planning," 544–52.
38. Fish and Wildlife Subcommittee, LAC Task Force, *Draft Supplement to Fish and Wildlife of the Bob Marshall Wilderness Complex and Surrounding Area Resource Management Organization,* Montana Department of Fish, Wildlife, and Parks (Helena, March 1989), 1.
39. B. Jennings, "Representation and Participation in Democratic Governance of Science and Technology," in M. Goggin, ed., *Governing Science and Technology,* 237.
40. U.S. Forest Service, *Shasta Costa: From a New Perspective* (Washington, D.C., 1990), 20.
41. A. Gillis, "The New Forestry," *BioScience* 40 (1990), 558–62.
42. U.S. Forest Service, *Shasta Costa,* 20.
43. U.S. Forest Service, *Draft Environmental Impact Statement: Shasta Costa Timber Sales and Integrated Resource Projects* (Gold Beach, Oregon, 1990), iv–53.
44. J. DeBonis, "New Perspectives: Editorial," *Inner Voice* 2: 4 (1990), 9.
45. S. Mathison and K. Wiedenmann, "Shasta Costa," *Forest Watch* 11: 4 (1990), 20.
46. A. Gillis, "The New Forestry," 560.
47. As quoted in R. Manning, "Fixing the Forests," *Buzzworm* 2: 6 (1990), 52.
48. R. Keiter, "Taking Account of the Ecosystem on the Public Domain," note 306 at 985.
49. Cascade Holistic Economic Consultants staff, "1991 Forest Service Bud-

get: More Incentives to Overcut the National Forests," *Forest Watch* 11: 4 (1990), S-1; M. Liverman, "Sustainable Timber Supply from Federal Forests in Oregon and Washington," Audubon Society of Portland memo, March 23, 1990. Available from author at Audubon Society of Portland, 5151 N.W. Cornell Road, Portland, Oregon 97210.

50. H. Rolston, "Property Rights and Endangered Species," *University of Colorado Law Review* 61: 2 (1990), 294–95.
51. M. Liverman, "Model Endangered Native Ecosystem Legislation, Third Draft," Audubon Society of Portland, February 20, 1990, p. 3.
52. See R. Noss and L. Harris, "Nodes, Networks, and MUM's"; C. Hunt, "Creating an Endangered Ecosystems Act," *Endangered Species Update* 6: 3–4 (1989), 1–5; Earth First! Biodiversity Task Force, "Constance Hunt's Article on Creating an Endangered Ecosystems Act," unpublished memo, May 9, 1989. Available from Jasper Carlton, Biodiversity Legal Foundation, P.O. Box 18327, Boulder, Colorado 80308.
53. M. Bader, "The Northern Rockies Ecosystem Protection Act: A Citizen Plan for Wildlands Management," *Western Wildlands* (summer 1991), 22–28.
54. M. O'Connell and R. Noss, "Managing Private Lands to Conserve Biodiversity," *Environmental Management* (in press).
55. D. Maehr, "The Florida Panther and Private Lands," *Conservation Biology* 4 (1990), 168.
56. B. Parry, "Cumulative Habitat Loss: Cracks in the Environmental Review Process," *Natural Areas Journal* 19 (1990), 77.
57. Ibid., 76.
58. Ibid., 77.
59. D. Jensen, M. Torn, and J. Harte, *In Our Own Hands: A Strategy for Conserving Biological Diversity in California* (Berkeley, 1990), 1–5.
60. California State Assembly Office of Research, *California 2000: Biological Ghettos* (Sacramento, 1991), 1–18.
61. Statement accompanying release of Memorandum of Understanding, "Agreement on Biological Diversity," California Resources Agency, Sacramento, September 19, 1991, p. 1.
62. T. Clark and J. Cragun, "Organization and Management of Endangered Species Programs," *Endangered Species Update* 8: 8 (1991), 3.
63. T. McNulty, "Olympic Peninsula Nightmare," *High Country News* (November 19, 1990), 12; K. Ervin, "The Politics of Compromise," *High Country News* (November 19, 1990), 15.
64. Erwin, "The Politics of Compromise," 15–16.
65. D. Foreman, "Reinhabitation, Biocentrism, and Self-Defense," *Earth First! Journal* (August 1, 1987), 22.
66. W. Berry, "Preserving Wilderness," in *Home Economics* (San Francisco, 1987), 139.
67. S. Hilts and T. Moull, *Natural Heritage and Stewardship Program Annual Report, 1986* (University of Guelph, Ontario, 1986), 20.

68. M. Van Patler, H. Geertz, and S. Hilts, "Enhancing Private Land Stewardship," *Natural Areas Journal* 10 (1990), 126.
69. A. Leopold, *A Sand County Almanac* (New York, 1949), 279.
70. A. Naess, "Intrinsic Value: Will the Defenders of Nature Please Rise?" in M. Soulé, ed., *Conservation Biology: The Science of Scarcity and Diversity* (Sunderland, Massachusetts, 1986), 504–15.
71. J. Carter et al., "The State of the Biology Major," *BioScience* 40 (1990), 680; for a treatment of ecological literacy in our culture, see D. Orr, *Ecological Literacy* (Albany, New York, 1992).
72. P. Ehrlich, "Facing the Habitability Crisis," *BioScience* 39 (1989), 480–82.
73. T. Fleischner and S. Weisberg, "Teaching for Nature," manuscript, 2. Available from primary author at Prescott College, 220 Grove Avenue, Prescott, Arizona 86301.
74. P. Feinsinger, "Professional Ecologists and the Education of Young Children," *Trends in Ecology and Evolution* 2: 2 (1987), 51–52.
75. "High-School Curriculum Ready for New School Year," *Update: Center for Conservation Biology Newsletter* 2: 2 (1988), 7.
76. C. Bohlen, "Biologists as Lobbyists," *BioScience* 40 (1990), 494. See also J. Warner et al., "Scientists and Journalists," *Bulletin of the Ecological Society of America* 72: 2 (1991), 116–18.
77. P. Morrison et al., *Ancient Forests in the Pacific Northwest* (Washington, D.C., 1991).
78. See D. Foreman, *Confessions of an Eco-Warrior* (New York, 1990).
79. There are numerous treatments of this topic. For accessible introductory works, see in general B. Barber, *Strong Democracy* (Berkeley, 1984); F. Lappe, *Rediscovering America's Values* (New York, 1989); G. Alperovitz, "Building a Living Democracy," *Sojourners* 19: 6 (1990), 10–23.
80. B. Barber, *Strong Democracy,* xiv.
81. D. Worster, *Rivers of Empire* (New York, 1985), 377.
82. U.S. Forest Service, "Review of and Comment on National Forest Plans and Project Decisions," *Federal Register* 57 (1992), 10445.
83. As quoted in J. Owens, "Report from Washington," *Western Ancient Forest Campaign Newsletter* 1: 20 (March 30, 1992), 11.
84. S. Bluestein, "Adopt-a-Wilderness '86," *Desert Notes* 5: 1 (1986), 1; D. Dagget, "Adoptions Used as Anti-Logging Tactic," *High Country News* (July 2, 1990), 7; a general account of grass-roots bioregional organizing is J. and C. Plant, *Putting Power in Its Place* (Philadelphia, 1992).
85. P. deLeon, "The STP Schools: Education for Environmental Action," *New Solutions* (summer 1990), 22–24.
86. K. Blanchard and M. Monroe, "Effective Educational Strategies for Reversing Population Declines in Seabirds," *Transactions of the North American Wildlife and Natural Resources Conference* 55 (in press).
87. D. Lewis, G. Kaweche, and A. Mwenya, "Wildlife Conservation Outside

Protected Areas: Lessons from an Experiment in Zambia," *Conservation Biology* 4 (1990), 171–80.

88. J. Gradwohl and R. Greenberg, *Saving the Tropical Forests* (Washington, D.C., 1988), 72–75.
89. R. Andrews, "Class Politics or Democratic Reform: Environmentalism and American Political Institutions," *Natural Resources Journal* 20 (1980), 221–41; D. Orr, "Reforming the Sierra Club," *Wild Earth* 1: 2 (1991), 36–37.
90. R. Barbee et al., "The Yellowstone Vision: An Experiment That Failed or a Vote for Posterity?" in R. Greenberg, ed., *Proceedings: Partnerships in Parks and Preservation Conference* (Washington, D.C., 1991).
91. B. Brown, "Notes on Social and Economic Transformations in Southwest Oregon," paper presented at the Poor Women in Timber Country Workshop, Northwest Women's Studies Association Meeting, Pullman, Washington, April 1991, 15–16.
92. R. Grossman, letter to author, December 9, 1991.
93. M. Sagoff, "Process or Product? Environmental Priorities in Environmental Management," *Environmental Ethics* 8 (1986), 136.
94. D. Foreman, "The New Conservation Movement," *Wild Earth* 1: 2 (1991), 6–12.
95. D. Johns, "The Relevance of Deep Ecology to the Third World: Some Preliminary Comments," *Environmental Ethics* 12 (1990), 239.

Chapter 7: Resources, Ecosystems, Place

1. J. Macy, "Bestiary," in J. Seed et al., *Thinking like a Mountain* (Philadelphia, 1988), 74–77.
2. See J. Seed et al., *Thinking like a Mountain,* 5–17, 97–113.
3. J. B. Callicott, "The Metaphysical Implications of Ecology," *Environmental Ethics* 8 (1986), 301.
4. For a summary see D. Foreman and H. Wolke, *The Big Outside* (Tucson, Arizona, 1989).
5. See A. Naess, "The Shallow and the Deep, Long-Range Ecology Movement," *Inquiry* 16 (1973), 95–100; see in general B. Devall and G. Sessions, *Deep Ecology* (Layton, Utah, 1985); W. Fox, *Towards a Transpersonal Ecology* (Boston, 1990).
6. S. Goldean, "The Principle of Extended Identity," *The Trumpeter* 3 (1986), 23.
7. A. Naess, "Self-Realization: An Ecological Approach to Being in the World," *The Trumpeter* 4 (1987), 36.
8. F. Matthews, "Conservation and Self-Realization: A Deep Ecology Perspective," *Environmental Ethics* 10 (1988), 355.
9. T. Birch, "The Incarceration of Wilderness: Wilderness Areas as Prisons," *Environmental Ethics* 12 (1990), 20.
10. B. Lopez, "The American Geographies," *Orion Nature Quarterly* 8: 4 (1989), 57.

11. W. Fox, "On the Interpretation of Naess's Central Term 'Self-realization,' " *The Trumpeter* 7 (1990), 99.
12. For recent views, see D. Botkin, *Discordant Harmonies* (New York, 1990); M. Pollan et al., "Only Man's Presence Can Save Nature," *Harpers* (April 1990), 37–48. For an older view, see R. Dubos, *The Wooing of the Earth* (New York, 1980).
13. W. Sachs and I. Illich, "A Critique of Ecology," *New Perspectives Quarterly* (spring 1989), 17–18.
14. B. Devall, *Simple in Means, Rich in Ends* (Layton, Utah, 1988), 66.
15. See A. McLaughlin, "Images and Ethics of Nature," *Environmental Ethics* 7 (1985), 293–320; T. Weiskel, "Agents of Empire: Steps Towards an Ecology of Imperialism," *Environmental Review* 11 (1987), 275–88.
16. H. Rolston, "Duties to Ecosystems," in J. B. Callicott, ed., *Companion to a Sand County Almanac* (Madison, Wisconsin, 1987), 258.
17. J. Livingston, "Some Reflections on Integrated Wildlife and Forest Management," *The Trumpeter* 3 (1986), 24–29.
18. M. Soulé, "Conservation Biology in the Twenty-first Century: Summary and Outlook," in D. Western and M. Pearl, eds., *Conservation for the Twenty-first Century* (New York, 1989), 303. See also M. Soulé, "The Onslaught of Alien Species, and Other Challenges in the Coming Decades," *Conservation Biology* 4 (1990), 234–35.
19. P. Passell, "The Cost of Limiting Greenhouse Effect," *San Francisco Chronicle*, November 20, 1989, p. 14. For the range of debate on this issue, see S. Schneider, *Global Warming* (New York, 1990), 298–303.
20. See. H. Rolston, "Biology and Philosophy in Yellowstone," *Biology and Philosophy* 5 (1990), 4–6.
21. J. Berger and J. Wehausen, "Consequences of a Mammalian Predator-Prey Disequilibrium in the Great Basin Desert," *Conservation Biology* 5: 2 (1991), 244–47.
22. J. Anderson, "A Conceptual Framework for Evaluating and Quantifying Naturalness," *Conservation Biology* 5: 3 (1991), 348.
23. A. Leopold, *A Sand County Almanac* (New York, 1949), 214.
24. N. Sakaki, *Break the Mirror* (San Francisco, 1987), 27.
25. R. Repetto, ed., *The Global Possible* (New Haven, 1985), 7; for a thorough review of the sustainable development concept, see D. Korten, "Sustainable Development," *World Policy Journal*, 9 (1991–92), 157–90.
26. G. Snyder, *The Practice of the Wild* (San Francisco, 1990), 92.
27. P. Berg and R. Dasmann, "Reinhabiting California," in P. Berg, ed., *Reinhabiting a Separate Country: A Bioregional Anthology of Northern California* (San Francisco, 1978), 218.
28. G. Esteva, "Regenerating People's Space," *Alternatives* 12 (1987), 133.
29. See W. Fox, *Towards a Transpersonal Ecology*, 124–25. For a recent collection of papers on various aspects of ecocentrism, see "From Anthropocentrism to Deep Ecology," a special issue of *ReVision* 13: 3 (1991).
30. I am indebted to Gary Snyder for the central ideas in the following section.

See his *Practice of the Wild,* 24–47. Also see M. Oelschlager, *The Idea of Wilderness,* 1–30.

31. B. McCay and J. Acheson, eds., *The Question of the Commons* (Tucson, Arizona, 1987), 23–25.
32. R. Nelson, *Make Prayers to the Raven* (Chicago, 1983), 246.
33. M. Zimmerman, "Quantum Theory, Intrinsic Value and Panentheism," *Environmental Ethics* 10 (1990), 29.
34. M. Oliver, *Dream Work* (New York, 1986), 85.

GLOSSARY

Ancient forest: the late-successional stages of forest development. Characteristics in western Cascades Douglas-fir forests include old trees (more than 200 years); large trees; a broken, uneven canopy; numerous snags; and fallen logs. Synonymous with old-growth forest.

Anthropocentrism: a view of life that places humans above all other species in value and importance.

Biodiversity crisis: the ongoing rapid loss of species and destruction of ecosystems.

Biodiversity-protection network: a system of integrated reserves designed to protect biodiversity at all levels into the future.

Biological assessment: an administrative document that reports on the status of given species in relation to a proposed federal action. Not as detailed as an EIS.

Biological corridor: a strip, swath, or other kind of functional habitat that allows animals and plants to move between otherwise isolated patches of habitat.

Biological diversity: the diversity of living things (species) and of life patterns and processes (ecosystem structures and functions). Includes genetic diversity, species and population diversity, ecosystem diversity, landscape and regional diversity, and biosphere diversity. Also known as *biodiversity.*

Biosphere: that part of Earth, from the depths of the ocean to the upper atmosphere, that supports life.

Biosphere reserve: a management model of the UN's Man and the Biosphere program where a core preserve is surrounded by buffer zones with use increasing away from the core.

Board foot: the standard measure for timber products. One board foot measures 1 foot by 1 foot by 1 inch thick.

Captive breeding: the practice of taking a species from the wild and placing it under controlled conditions to facilitate successful reproduction.

Census population: a gross count of all individuals of a given species in a certain area.

Clearcutting: a logging method by which an entire forest stand is cut down. In national forests, the size of an individual clearcut cannot exceed 40 acres.

Commons: large areas of joint-use land with social control resting in the hands of local communities and standards of use built on intimate knowledge of plants, animals, and ecosystems. Most commons were destroyed with the advent of industrial civilization, but some still exist in less developed parts of the world.

Conservation: as originally coined by Gifford Pinchot, the development of natural resources for "the greatest happiness for the greatest number [of humans] over the longest period of time." Aldo Leopold defined conservation as "a state of harmony between people and land." Conservation biologists see it in terms of the long-term viability of species.

Conservation biology: the field of biology that studies the dynamics of diversity, scarcity, and extinction.

Council of All Beings: a group ceremony in which participants bear witness to the biodiversity crisis and seek empowerment to work in defense of wild nature.

Critical habitat: under the Endangered Species Act, "habitat essential to the conservation of species."

Cumulative-effects analysis: as required by the National Environmental Policy Act and similar state laws, analysis of the effects of planned land-management activities in a given area into the foreseeable future.

Deep ecology: a movement whose members seek, through ecological understanding, advocacy, and direct encounter with nature, an expanded and deepened relationship with the natural world. Deep ecologists seek an authentic alternative to anthropocentrism.

Deme: a local unit of a larger population.

Demography: the quantitative study of populations with reference to size, density, growth, distribution, etc.

Disturbance: in ecosystems, an event that interrupts succession, eliminates some part of the existing plant and animal community, and creates conditions for renewed growth and colonization. Examples are wildfire, windstorm, flooding, insect outbreaks, etc.

Ecocentrism: the belief that people are part and parcel of nature, not superior to it. Ecocentrism maintains that a main task of humans is to learn how to fit in with natural ecosystems; its methods are cooperative instead of competitive. It is not misanthropic but rather seeks ways to value all beings in the world without devaluing humans.

Ecological self: a self beyond the individual or social self. A person with a developed ecological self experiences a profound connection with nonhuman life forms, mountains, rivers, etc.

Ecosystem: a community of species and its physical environment. When defined at different levels, it often involves arbitrary boundaries. An ecosystem may refer to anything from a fallen log to an entire watershed.

Ecosystem diversity: the variety of ecosystems within a given area.

Ecosystem functions: the different physical and chemical processes of ecosystems, that is, the water cycle, the nutrient cycle, the breakdown of living matter (decay), etc.

Ecosystem management: any land-management system that seeks to protect viable populations of all native species, perpetuate natural-disturbance regimes on the regional scale, adopt a planning timeline of centuries, and allow human use at levels that do not result in long-term ecological degradation.

Ecosystem structure: the physical patterns of life forms within an ecosystem. In a Douglas-fir forest, for example, there is a great amount of vertical structure from the forest floor to the top of the canopy. A desert-shrub ecosystem has a simple physical structure.

Edge effects: the ecological changes that occur at the boundaries of ecosystems. These include changes in species composition, gradients of moisture, sunlight, soil and air temperature, wind speed, etc. Not all edge effects are positive; many forest-interior species have their populations reduced by edge effects.

Effective population: the number of individuals in a given group that may be expected to contribute reproductively to the next generation.

Emergent properties: a part of hierarchy theory that suggests that what happens at higher levels (the biosphere) constrains lower levels (ecosystems).

Endangered: a legal classification of the federal Endangered Species Act under which a species is at risk of becoming extinct throughout all or a significant portion of its range.

Environmental impact statement (EIS): a review of the effects of a major federal land-management proposal as required by the National Environmental Policy Act.

Forest-interior species: species that are adapted to conditions deep within unfragmented forest stands.

Forest plan: the comprehensive land-management plan required of each national forest under the National Forest Management Act.

Forest plan appeal: a formal request for an administrative ruling on a contentious issue in a forest plan. The appeal process must be exhausted before a legal suit may be brought.

FORPLAN: the computer model used by the Forest Service in forest planning.

Gap analysis: a method of identifying important areas of biodiversity that remain unprotected.

Genetic diversity: the amount of genetic variation among organisms in a population and among the populations of given species; the lowest level of biodiversity.

Genetic drift: changes in the genetic diversity of a given population due to random events over the long term.

Global warming: the projected increase in global temperature due to the release of the byproducts of fossil-fuel combustion into Earth's atmosphere.

Greater ecosystem: a regional complex of ecosystems with common landscape-level characteristics such as the presence of wide-ranging mammal populations and periodic, large, natural disturbances such as wildfire. The concept of the greater ecosystem allows for integrated land management. The boundaries of greater ecosystems are somewhat arbitrary.

Habitat fragmentation: destruction of habitat through loss of functional habitat and the isolation of the remaining patches within an ecosystem.

Heterozygosity: having the genes on paired chromosomes different at one or more locations. Increases genetic diversity within a population.

Hierarchy theory: a theory of relations among different levels (genes, species, etc.) of biodiversity. Suggests that each level has unique independent properties as well as properties that function as part of a whole.

Interagency Grizzly Bear Committee (*IGBC*): a standing committee of various federal, state, and Indian agencies focused on management of the grizzly bear.

Island biogeography: the study of the distribution of living things, especially on islands.

Keystone species: a species that plays a role in an ecosystem that far outweighs the role of other species.

Landscape diversity: patterns that link different kinds of ecosystems across large regions. Includes the effects of large-scale natural disturbances such as fire, the interactions of widely dispersed populations that may use biological corridors, etc.

Management indicator species (*MIS*): a species that theoretically indicates the general health of an ecosystem. A decline in its population signals decline for the rest of the species living in the area.

Metapopulation: a group of subpopulations of one species where the movement among groups is much less than movement within an individual group. A good example of this is the grizzly bear in the continental United States.

Minimum dynamic area: the smallest area subject to a pattern of natural disturbance that retains an internal recolonization source. The idea is that a nature reserve must be at least this size to prevent a disturbance from so altering conditions that local species go extinct.

Multiple-use: the dominant land-management paradigm of the U.S. Forest Service. It mandates the use of forests for a variety of resources and activities, including timber, watershed protection, recreation, and wildlife range.

National forest: public lands administered by the U.S. Forest Service. There are 156 national forests, totaling about 191 million acres of land.

National park: lands administered by the National Park Service primarily for their natural beauty and recreational offerings. Not all Park Service lands are designated as national park; there are also monuments, historical sites, etc.

Native diversity: the diversity of species that have evolved in a given place without the influence of humans.

Natural ecosystem: an ecosystem whose species and ecological structure and function remain largely unaffected by human activity. *Natural* is a relative term; see the discussion in chapter 7.

Nature reserve: any natural area that is relatively well protected from degradation, for example, national parks and wilderness.

New Conservation Movement: a growing movement of grass-roots environmental advocates who base their activism in local and regional issues, ethical obligations to ecosystems, and an ecosystem view of nature.

New forestry: a type of forest management that seeks to incorporate ideas from conservation biology and landscape ecology into timber management.

New Perspectives: an experimental program of the Forest Service that purports to incorporate ecosystem-management principles and increased public participation into forest planning.

Old-growth forest: See Ancient forest.

Paleoecology: the study of prehistoric patterns of species diversity and distribution.

Population: the interbreeding individuals of a species inhabiting a given area.

Population-viability analysis: a system of modeling the synergistic effects of genetic, demographic, and environmental uncertainty on a given population.

Preservation: the designation of parks and wilderness areas prior to the rise of an ecosystem view. This strategic goal of the wilderness movement derives from the nineteenth-century romantic urge to find balance between wilderness and civilization.

Presettlement vegetation: the vegetation that existed in America prior to the coming of Euro-Americans.

Public lands: those lands owned by the American people and managed by the U.S. government.

Relaxation rate: the rate of the continuing loss of species in a given location after the area has been cut off from genetic exchange with neighboring areas.

Resource: a product of the natural world that is useful to humans.

Resourcism: a view of the world that considers nonhuman beings to have value only as they can be used as goods and services for humans.

Restoration: the recreation of communities of organisms resembling those occurring prior to degradation.

Snag: a standing dead tree.

Species: a group of organisms that (usually) can reproduce successfully only with each other.

Species diversity: the variety of species inhabiting a given area.

Species recovery: the approach to maintaining the viability of a species formerly threatened with extinction.

Stand-replacement fire: a fire that completely removes the forest over a given area.

Stochasticity: uncertainty due to random events. In conservation biology, this includes genetic, demographic, and environmental processes.

Succession: change in the structure and composition of ecosystems over time.

Sustainable: describes levels of human use that allow ecosystems to retain their basic structure and function over the long term.

Sustained yield: the amount of a resource taken in perpetuity without damage to the resource base.

Taking: under the Endangered Species Act, any action to "harass, harm, pursue, hunt, shoot, wound, kill, capture, etc." a listed species.

Threatened: a legal classification under the federal Endangered Species Act that describes a species as likely to become endangered in the foreseeable future.

Utilitarian: in land management, the concept that natural resources exist for human use. Gifford Pinchot called this wise-use.

Viable population: a population that stands an excellent chance of surviving with minimal human management.

Wilderness: as defined by federal law, any area where the imprint of humans is substantially unnoticeable and natural processes operate relatively freely. Many areas meet the legal definition but are not formally classified as wilderness.

INDEX

ABOUT THE AUTHOR

Since 1982, R. Edward Grumbine has been director of the Sierra Institute, a wildlands studies undergraduate program of the University of California Extension, Santa Cruz. He has taught courses on field ecology and ecosystem management in outdoor classrooms throughout the western United States. Dr. Grumbine is currently working on translating conservation biology concepts and ecocentric values into federal land management policies and practices. He lives in Bonny Doon, California.

ALSO AVAILABLE FROM ISLAND PRESS

Balancing on the Brink of Extinction: The Endangered Species Act and Lessons for the Future
Edited by Kathryn A. Kohm

Better Trout Habitat: A Guide to Stream Restoration and Management
By Christopher J. Hunter

Beyond 40 Percent: Record-Setting Recycling and Composting Programs
The Institute for Local Self-Reliance

Coastal Alert: Ecosystems, Energy, and Offshore Oil Drilling
By Dwight Holing

The Complete Guide to Environmental Careers
The CEIP Fund

Death in the Marsh
By Tom Harris

Farming in Nature's Image
By Judith Soule and Jon Piper

The Global Citizen
By Donella Meadows

Healthy Homes, Healthy Kids
By Joyce Schoemaker and Charity Vitale

Holistic Resource Management
By Allan Savory

Inside the Environmental Movement: Meeting the Leadership Challenge
By Donald Snow

Last Animals at the Zoo: How Mass Extinction Can Be Stopped
By Colin Tudge

Learning to Listen to the Land
Edited by Bill Willers

Lessons from Nature: Learning to Live Sustainably on the Earth
By Daniel D. Chiras

The Living Ocean: Understanding and Protecting Marine Biodiversity
By Boyce Thorne-Miller and John G. Catena

Making Things Happen
By Joan Wolfe

Media and the Environment
Edited by Craig LaMay and Everette E. Dennis

Nature Tourism: Managing for the Environment
Edited by Tensie Whelan

The New York Environment Book
By Eric A. Goldstein and Mark A. Izeman

Our Country, The Planet: Forging a Partnership for Survival
By Shridath Ramphal

Overtapped Oasis: Reform or Revolution for Western Water
By Marc Reisner and Sarah Bates

Plastics: America's Packaging Dilemma
By Nancy Wolf and Ellen Feldman

Race to Save the Tropics: Ecology and Economics for a Sustainable Future
Edited by Robert Goodland

Rain Forest in Your Kitchen: The Hidden Connection Between Extinction and Your Supermarket
By Martin Teitel

The Rising Tide: Global Warming and World Sea Levels
By Lynne T. Edgerton

The Snake River: Window to the West
By Tim Palmer

Steady-State Economics: Second Edition with New Essays
By Herman E. Daly

Taking Out the Trash: A No-Nonsense Guide to Recycling
By Jennifer Carless

Trees, Why Do You Wait?
By Richard Critchfield

Turning the Tide: Saving the Chesapeake Bay
By Tom Horton and William M. Eichbaum

War on Waste: Can America Win Its Battle With Garbage?
By Louis Blumberg and Robert Gottlieb

Western Water Made Simple
From *High Country News*

For a complete catalog of Island Press publications, please write:
Island Press, Box 7, Covelo, Ca 95428, or call: 1-800-828-1302

ISLAND PRESS
BOARD OF DIRECTORS

SUSAN E. SECHLER, CHAIR
Director
Rural Economic Policy Program
Aspen Institute for Humanistic Studies

HENRY REATH, VICE-CHAIR
President
Collector's Reprints, Inc.

DRUMMOND PIKE, SECRETARY
President
The Tides Foundation

PAIGE K. MACDONALD, ASSISTANT SECRETARY
Executive Vice President/Chief Operating Officer
World Wildlife Fund & The Conservation Foundation

GEORGE T. FRAMPTON, JR., TREASURER
President
The Wilderness Society

ROBERT E. BAENSCH
Senior Vice President/Marketing
Rizzoli International Publications, Inc.

PETER R. BORRELLI
Vice President of Land Preservation
Open Space Institute

CATHERINE M. CONOVER

PAUL HAWKEN
Chief Executive Officer
Smith & Hawken

CHARLES C. SAVITT
President
Center for Resource Economics/Island Press

PETER R. STEIN
Managing Partner
Lyme Timber Company

RICHARD TRUDELL
Executive Director
American Indian Resources Institute